Vinyl
Sign Techniques

Jim Hingst

MEDIA
GROUP
INTERNATIONAL

Cincinnati, Ohio

This book has been compiled from 54 "Vinyl Graphics" columns by Jim Hingst for *Signs of the Times*, between July, 2001 and February, 2007. Every column was checked and updated by the author in January, 2007. The two-part columns were combined into single chapters, which is how 54 columns became 39 chapters of this book.

ISBN: 0-944094-53-8

Published by:

ST Media Group International Inc.
Book Division
407 Gilbert Avenue
Cincinnati, Ohio 45202
USA

Tel. 800-825-1110 or 513-421-2050
Fax 513-421-6110
E-mail: books@stmediagroup.com
Online bookstore: www.stmediagroup.com

Book design: Jeff Russ, Senior Art Director, *Signs of the Times*

Cover photos courtesy of Avery Dennison, Blue Media and R Tape.

Printed in the United States of America

10 9 8 7 6 5 4 3 2   First Printing

Acknowledgments

I would like to thank all of my friends in the sign and screen print industries, who have helped me through the years on my articles and on this project. These individuals have provided me with the information that I needed, and have patiently read through my rough drafts, offering their editorial criticism. I am also grateful to several of my colleagues at R Tape Corporation and Custom Extrusion Technologies, who have contributed their expertise and thoughts on the subjects that I cover in this book. The following is a partial list of those have helped with my research:

Chuck Bules of Arlon

Tim Doyle, Dave Harris, Alan Weinstein
and Molly Waters of Avery Dennison

Judy Eck and Tom Cornelius
of FDC Graphic Films

Bea Purcell and Laure Maybaum of Nazdar

Butch "SuperFrog" Anton
of SuperFrog Signs & Graphics

Laura Wilson of Roland

Karel Blonde and Jo Vanneste of Industrial Consultants
(R Tape Europe)

Greg McKay and Joe Balabuszko
of Earl Mich Company

Rich Thompson of Ad Graphics

Bill Barnes of Falcon Enterprises, Inc.

Tom Pidgeon and Bill Hammann
of D&K Group, Inc.

John Gooch of American Renolit

Jenifer Codrea and Mike Popovich of Kapco

I would also like to thank my friends in the *Signs of the Times* organization for their support and help. Many thanks to Wade Swormstedt for giving me the opportunity to contribute to the "Vinyl Graphics" column. Other people, who I wish to thank for their editorial assistance with this book, are Susan Conner, Steve Aust and Shannon Reinert, as well as Jeff Russ for his work in the design and layout of this book. I am also grateful to Nancy Bottoms and Laura Leonhartsberger for translating many of my columns into Spanish and giving me the space in *Signs of the Times & Screen Printing en español*. Finally, my articles may never have been compiled as a book had I not received encouragement and support from Mark Kissling.

Contents

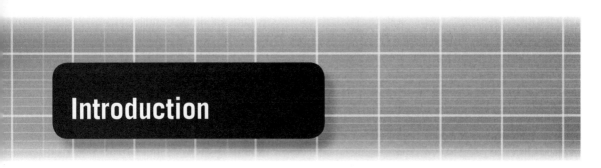

Introduction

Why did I write this book?

A little over twenty-five years ago, I started working for a large fleet graphics screen printer in northwest Indiana. I was fortunate enough to undergo a month-long indoctrination and training program, during which I had the opportunity to work in all aspects of the business. I spent time with the estimators, production planners, stencil cutters and creative artists. Then, I worked a couple of weeks in the shop. During the final week of training, I learned how to apply vinyl at one of the trailer manufacturers. At the end of my first month, the general manager of the company gave me a binder, filled with articles on fleet graphics and technical bulletins. I was told to study it every day.

While that may not be much of an education, it amounts to more training than what most get in the industry. After the indoctrination period, I was on my own.

I received some guidance from some the seasoned veterans, but most of what I learned was through trial and error – lots of errors.

I was lucky that I had a few years of work experience in the offset printing industry and a formal education in drawing and design and the graphic arts. My background helped me make the transition into screenprinting and vinyl graphics. Many others weren't so lucky and didn't make the cut. Their failures were not the result of a lack of trying; instead, a lack of training was the cause.

After I had a few years of experience under my belt, two friends, Tom Barnard and Paul Godfrey, started a screenprint company, Signature Graphics in Porter, Indiana, and gave me the chance to work on the production side of the business. Again, most of what I learned was trial and error – much of it at the expense of my friends. During this period, I had the opportunity to start writing a formal training program, intended to give new employees and manufacturer's reps an overview of the business.

As I moved from one company to another, I expanded that training program as I learned more about the business. After a number of revisions, my writings eventually evolved into 53 articles that I wrote for *Signs of the Times*. Ultimately, my goal was to put all these articles together in a comprehensive book about vinyl graphics.

Who is this book for?

This book is appropriate for anyone who is seriously involved in the vinyl graphics business, whether you are working for a sign company, screen printer or a materials manufacturer.

If you are new to the business, this book will provide you with ideas to sell and market graphics programs. It will also give you a comprehensive education on the fundamentals of the materials and techniques used to manufacture, install and remove vinyl graphics.

If you are an owner or manager of a business, this book can provide you with the information that you need to coach your people so you can supplement their understanding of the business and improve their performance in your organization.

What's in this book?

This book is based on thirty-plus years of learning experience in the graphic arts field. It is filled with real life examples of selling, designing, manufacturing and installing vinyl graphics.

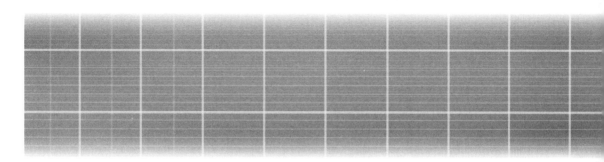

The book is organized into four major sections: sales and marketing, materials, fabrication, and installation and removal. I thought that sales and marketing was a good place to start, because nothing happens in business until the sale is made. I learned this from the person who recommended me for my first screenprinting sales job, Davy Crockett. (No, not the guy in the coonskin cap, who fought and died at the Alamo – I'm old but not that old).

Davy Crockett had been a sales trainer for 3M and had written the book, *For Those Who Sell, and Who The Hell Doesn't*. The central theme of the book was that whether you are a receptionist, designer, or production person, your most important job is to sell and promote your company's products and services. This is the job of everyone in the business. Sales, not quality, is job #1.

Section two of the book cover materials. Although I have worked for a few companies that manufacture pressure-sensitive vinyl, and a couple of companies that manufacture application tape, none of these companies formally train their employees on adhesives and films, beyond conducting a plant tour and a few talks at sales meetings.

I first learned about the chemistry of vinyl and the sticky stuff from a chemist at Arlon, Howard Anderson. During a business trip in Canada, Howard and I had a free day. That morning I asked him if he would give me some product training. Howard talked for eleven hours straight. As he talked, I listened and wrote. Many other friends have taken the time to answer my questions. Much of what I learned from these friends is covered in the chapters on vinyls and adhesives.

The section on materials will give you a good understanding of which materials to use, how these materials are made and their performance characteristics. These chapters review a lot of theory. The remaining two sections cover the practical aspects of graphics fabrication, vinyl application and film removal. Some of the chapters explain processes and techniques that you may never have tried, such as wrapping a vehicle or doming or screenprinting. This book provides the explanation.

To really understand any of the concepts explained in this book or any other, you need to put the ideas into action. Take Nike's advice and "Just do it". For the last few years, I had been conducting regular training sessions with our customer service reps, in which I was the performer and they were spectators. They finally complained that they wanted hands on training. They wanted to be participants in their own education.

Consequently, I have changed the format of our training. In recent sessions, the R Tape customer service people have been applying vinyl, painting using our paint mask vinyl and most recently screenprinting. In the screenprinting training session, they did everything from taping the screens to pulling a squeegee. Within two hours, our reps knew the basics. They also learned that screenprinting can be a messy business, so they left the clean up to me.

If you want to learn the fundamentals of vinyl graphics, please read on and then "just do it." Otherwise, this book will make a fine paperweight or a Christmas gift for a friend in the industry.

— *Jim Hingst, January 2007*

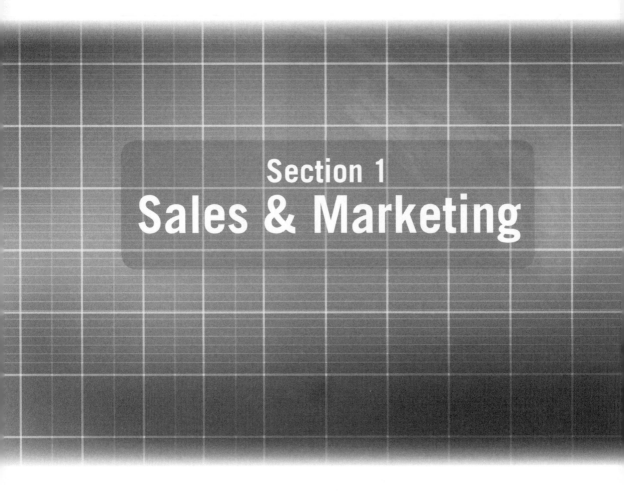

Section 1
Sales & Marketing

Chapter 1
Sales and Marketing Basics

Approximately five years ago, I visited a Louisville, KY-based signshop owner. After unsuccessfully competing for several, large, fleet-graphics jobs, he said that, not only was he losing every bid, he wasn't even in the game. In many cases, winning selling prices were lower than his raw-material costs.

That really shouldn't come as a surprise — big companies have tremendous buying power. Why waste your time on jobs that you're unlikely to win? Despite biblical stories, the smart money is still on Goliath.

To motivated vinyl-graphics fabricators and salespeople, these rolls of vinyl represent dollar bills. Get material off the racks and onto vehicles or storefronts — and cashflow into your pocket — by proactively locating and pursuing prospective clients, and implementing software and other tools to support a related database.

A former boss, Gordon McAllister, advised me that, rather than trying to score the big sales, concentrate on discovering those opportunities you and your competitors are neglecting.

Work the phones

Unearthing these sales gems requires prospecting — investing time on your telephone, PC or laptop. Your goal should be to build a base of customers willing to pay your price.

Existing customers could provide some business opportunities you routinely miss. Have you really investigated all of your customers' graphics applications? If you're selling a client fleet graphics, could you augment your sales by also offering window or interior graphics?

A lucky few have made fortunes overnight. Get-rich-quick schemes usually crash and burn, and big sales that net financial windfalls seldom develop. People I know in our industry, who own luxurious homes and drive exotic cars, didn't hit the jackpot with one big sale. They made their money by working longer, harder and smarter than their competitors.

There's no easy way to generate sales. If there is, I haven't discovered it. In this article, I'll outline a step-by-step process that many graphics providers and salespeople have used successfully. This plan involves setting personal and company sales goals; prospecting to unearth opportunities; discovering who makes decisions for your potential customers; determining how you can satisfy the prospect's needs; learning how to professionally state your case; and providing the service needed to maintain business.

That's quite a list, and it entails a great deal of work. But by following this step-by-step approach, you can influence the outcome of the sale and control your own destiny and fortune.

Due diligence

My perfect salesperson was, on the surface, an unlikely candidate. Jerry was short and thin (about 135 lbs. soaking wet) and didn't have a college degree. What's more, he had no previous sign-industry experience — I think he trimmed trees before becoming a salesman — and was humble and soft spoken.

I probably wouldn't have hired him. Yet, beneath his exterior, Jerry is intelligent, tenacious and courageous. Most of all, Jerry worked hard; five days a week, 50 weeks a year. Salesmen like that are rare.

Unlike Jerry, most salespeople I know create excuses not to work. Mondays are filled with expense reports and paperwork. Midweek, sales take a back seat to running errands. Fridays, they wouldn't dare disturb a prospect who's pondering weekend plans.

In contrast, Jerry handled his paperwork on the weekends. During the week, he was on the road when his competitors were just waking up. By 8 a.m., Jerry was outside the door of his first prospect. He averaged 12 sales calls per day.

Jerry didn't usually make the big sale. Instead, he made numerous little sales that averaged $5,000 per day, at 20% commission. His annual revenue always topped $1 million, making him more than $200,000. Hurray for the Jerry's of this world! They deserve every penny.

Successful selling is more than just a numbers game. But, if you don't step up to the plate and swing the bat, the hits won't come. Roger Jacobs, CEO of R Tape (South Plainfield, NJ) is fond of saying, "A funny thing happens when you go out and make a sales call. You start making sales."

A tale of two salesmen

Although some people are better suited for sales, I don't believe that salespeople are "born." Fifteen years ago, I hired and trained two very different individuals, whose careers took divergent paths.

Ed was a "born" salesman; he was tall, handsome and charismatic, and he dressed the part. My boss thought he looked like a Harvard Business School graduate and told me to hire him.

Terry, on the other hand, wasn't blessed with the same natural gifts. He was overweight and not particularly charming. However, he was detail-oriented and possessed time-management skills. Also, he showed undaunted determination to achieve his goals. If I told him to do something, he wrote down the task and accomplished it without fail.

A devoted husband and

Jim's Tips

- All the talent in the world won't overcome a lack of hard work.
- If you hire salespeople, pay them on commission.
- Telemarketing is a valuable source for information, prior to making sales calls.
- Use database software, such as ACT, to track existing customers and prospects.
- Create direct-mail campaigns that will grab potential customers' attention. If you see graphic problems on their vehicles or in their stores, take pictures and send them with the literature.

father, Terry felt responsible for providing a higher standard of living for his family. What could be a better goal? Today, Terry remains a successful and respected salesperson with the same graphics company.

So what happened to Ed, the "natural"? After some initial successes, he wasted his time on moneymaking schemes. He had dreams of financial wealth, but he didn't want to work.

Like many other glib salesmen, Ed tried to wing everything. With no direction, he drifted from one job to another. Instead of working, he partied. His marriage ended in divorce, and, after a couple DUI convictions, he lost his driver's license. The last I heard, he was working odd jobs as a handyman. The world is full of talented people like Ed who never amount to anything.

Success in sales, or anything else for that matter, takes more than talent. You need to set goals and develop plans to achieve them. Further, you need to establish a timetable for putting those plans into action, and then take action. If you don't do these things, there's no hope for you.

Money motivated

The people most successful at selling fleet and building graphics have been obsessed with accumulating financial wealth. These people dedicate themselves to learning everything they need to know about the business to achieve their dreams. They have exceptional knowledge of design, raw materials, manufacturing processes and installation procedures. And they pursue their dreams with persistence, passion and an unshakable confidence that they will succeed at any cost.

The vinyl-graphics industry, like any other, centers on making money. It's not about making the most aesthetically pleasing design or working harder for job satisfaction. The best salespeople realize that money is the best motivator.

This is why, historically, top salespeople work on commission. The equation is simple: If you make the sale, you get paid; if you don't, you won't. When sales equate to feast or famine for themselves and their families, salespeople make sales with a greater sense of urgency, intensity and commitment.

If you need to hire a salesperson, hire the hungry ones that will seek and devour opportunities with avarice. Give a real salesperson the opportunity to make money, and he will take care of himself and your business.

If your salaried salesperson isn't giving you a good investment, terminate him or put him on straight commission. In most cases, firing is your best option. Trying to motivate an unproductive employee, with poor work habits or weak character, is usually futile. More importantly, keeping a poor performer often demoralizes other team members.

Marketing 101

A popular adage says, "Success depends on planning your work, and then working your plan." In many organizations, the plan becomes a formal marketing strategy. This plan should simply and clearly establish your sales goals and outline the action steps required to achieve them.

The plan must also define marketing and sales activities in specific, easily understood and measurable terms. More importantly, everybody involved needs to read and under-

When proposing a vehicle-graphics job to a prospective client, communicate your proposal to all major parties. The fleet manager typically schedules the trucks for installation; the advertising or marketing manager controls the company's corporate image, so the job could come out of his budget; and, the president or owner will have final say over any sizeable purchase because he's in charge of profit and loss.

stand the company goals, the activities required to achieve them, the timetable for accomplishing assigned tasks and each individual's responsibilities as part of the team.

To help motivate employees, explain why attaining individual goals is critical to your organization's success. Most importantly, put the plan into action and review the plan regularly so everybody stays on track.

To monitor such a program, maintain weekly marketing and sales records. This activity report should include the number of records added and deleted to your sales database; the number of qualifying phone calls made; the number of direct-mail packages sent; advertising inquiries; the number of sales calls; and quotations and orders.

In addition to a report, rolling sales forecasts should be maintained. A sales forecast is an excellent way to record opportunities and monitor salespeople's performance. A forecast also helps prevent missed opportunities.

Each prospect should be listed in the rolling forecast. This forecast can be easily formatted and maintained as a spreadsheet file. The file should list fields for the account name, annual revenue potential, the probability of getting the business and each month's results. Multiply the potential by the probability, and enter the projected sales amount in the month that the prospect is likely to make his buying decision. Review the forecast and the sales-activity report with key members of your team each week. It will continually remind you of what you've done and need to do.

Prospecting

Telemarketing can economically unearth new customers and protect existing ones. Telemarketing is no substitute for face-to-face selling, but it's certainly an effective tool for uncovering opportunities.

Telemarketing qualified potential clients helps determine: which companies operate, for instance, a fleet of vehicles or a chain of stores with an existing graphics program; who has purchasing authority within these companies; and when they are likely to purchase these products.

Software

Accurate recordkeeping is critical for telemarketing and sales. If you aren't using a contact database program now, look into one. Programs such as ACT are reasonably priced and popular with many salespeople. With Microsoft Outlook or Access, you may not need a special program.

One of our salesmen at R Tape has used Outlook since he started with the company. With it, he can print mailing labels for direct mail, and send blast faxes and mass e-mailings. He can also schedule any follow-up activities, such as sending direct mail or arranging a sales call.

Another plus is the program's simplicity, so there's virtually no learning curve for anyone with even minimal computer skills. Above all, use some type of program.

In the software's detail screen, list any pertinent information about your competitors, especially those currently providing the graphics to your prospect. In your notes, include the competitor's strengths and weaknesses. Also note any changes that might be occurring within the account, such as reorganization, or changes in titles or duties, that could potentially help or hinder your prospects for future business.

In the profile of a vehicle-graphics customer, include the size of the prospect's fleet, type of vehicles, and business objectives and marketing themes. Also, keep notes regarding customers' personal information, such as education, business background, family, interests and hobbies.

Does the prospect feel obligated to his existing supplier? Understanding your prospect as a person changes the playing field from a sterile business atmosphere to a personal relationship. Remember, a key to selling is creating a climate in which the prospect feels comfortable and confident doing business with you.

Of course, there's nothing wrong with selling to a friend. As one trailer dealer once told me, "Make money from your friends, because you're sure not going to make any from your enemies."

Direct mail

After the prospecting phone call, follow up with, at the very least, a letter. This reinforces your telemarketing efforts. Include printed literature that shows the quality of your work. You might even include some of the vinyl manufacturers' brochures, explaining the value of graphics.

When developing direct-mail materials, understand your audience. To whom are you selling, and what do they really want? You can structure your marketing message in various ways. Appealing to potential clients' egos — and, in turn, their sense of pride in how their business is represented, including a graphic program — is one option. Or, focus your message on the public relations and advertising value of effective commercial graphics.

Advertising, direct mail, newsletters or even a short letter of introduction helps open many doors to decisionmakers who would otherwise be inaccessible. Marketing should create a favorable climate that makes prospects receptive to hearing your sales message. Regularly sending newsletters or direct mail, for example, helps build your company's name recognition and credibility. It also acquaints would-be customers with your products and services. Plus, direct-marketing materials help prospects recognize your name and company when you call for an appointment.

Direct mail should emphasize the primary benefits prospects will derive from doing business with your company. The message needs to provide the decisionmaker a reason to give you some of his valuable time to listen to your presentation. What value do you offer? What makes your sales proposal unique and different? What sets you apart from your competition? What are your company's strengths, and how can you leverage them to your advantage?

Before writing your letter of introduction and calling prospects, collect information about them. During this intelligence gathering, try to discover why they might want to change their graphics vendor. Remember, if prospects are content with their current program, you really don't have an opportunity.

I worked for one very successful sales manager who always sent prospects a letter before calling.

If he noticed a graphics problem, such as vinyl that's edge-lifting, cracking, or fading, he would photograph the failure and send the picture with his letter. In the letter, he would merely state that he wanted to call this problem to their attention and would like an opportunity to discuss ideas for preventing recurring problems.

A few days after sending the letter, he would follow up with a call. Using a letter with photographs was his technique for gaining prospects' attention and disrupting their homeostasis. If you don't shake things up a little, there's no reason for potential customers to change the status quo.

Of course, not everyone we target has an acute problem, so we can't always save the day with innovative graphic solutions. Many of our prospects are perfectly satisfied with their existing graphics programs. In cases such as these, you might consider sending prospects a photograph of a competitor's new graphics. Nobody likes to be outdone by a rival company, so the picture might generate interest and give them a reason to want to talk to you.

Getting Past The Gatekeeper

A few years ago, a signmaker explained her problems getting past receptionists, which prevented her from speaking with prospects. Regardless of who you are, getting past the "gatekeeper" is a difficult hurdle.

Receptionists and secretaries are often instructed to screen their boss's calls. Many such veterans have heard every sales pitch imaginable, and you'll rarely outfox them. Intimidating or bullying a receptionist doesn't work either; if you push, be prepared for that person to push back.

Rather than treating gatekeepers as adversaries, treat them with respect. Often, they don't get the credit they deserve on the job. Try to make them feel important.

One salesman I know introduced himself and said, "I need your help. I design and manufacture truck graphics. To whom should I speak?" He followed by asking for a convenient time to call or see that person. Open-ended questions helped make this person feel comfortable.

In many other business-to-business sales, gatekeeper interaction requires time and effort to understand the intricacies of the prospect's internal structure and buying process. Telemarketing yields some answers, but you can't build relationships with a phone call. Nothing replaces face-to-face meetings.

Making the call

Within large companies, many players usually influence purchasing decisions. During graphics sales, the president, advertising manager, marketing manager, fleet manager and purchasing manager could all be involved.

Fleet managers often schedule the trucks for installation, and they manage the program. The advertising or marketing manager is responsible for maintaining corporate image, and the money could come out of his/her budget. Purchasing managers act as gatekeepers to screen out unqualified vendors. And, of course, the company owner or president exercises final say.

So, you need to understand company dynamics. You'll often need to state your case several times to those with influence. If you don't pull all key people into the process, someone may sabotage your efforts. Although the company president may like your proposal, others within the organization can kill the deal. These individuals typically implement major graphics programs. Many prefer the status quo; it means less work for them.

To understand an organization's complexities, try to cultivate a guide or sponsor. This person can help you navigate potential minefields. Your sponsor may be someone with whom you've previously done business. This ally can help you identify decisionmakers, avoid time-wasters, alert salespeople to problems, and help track progress during the selling process.

When pitching a sale to a company, it's key to have either an ally within the company or another vendor who works closely with the company. Sometimes a simple gesture such as buying them coffee or breakfast can create goodwill and help open doors.

Your sponsor mustn't necessarily work for the prospect. For example, if you're selling fleet graphics, you can glean valuable information and guidance from leasing-company salespeople. They know whom to contact, as well as who's buying what and when. A former sales manager encouraged me to start each morning by taking a leasing salesperson out for breakfast or coffee.

In many cases, industry friends will gladly recommend you to companies looking for a dependable, professional graphics provider. A network of contacts can inform you when changes occur, such as employees changing positions, or when a company buys a fleet of trucks or builds new stores.

A business' graphics purchases are usually an ongoing process, involving many different components, decisionmakers and contracts up for bid. Harvesting all account opportunities requires persistent, close contact. Identify individual opportunities and target each as its own sale, and develop an individual strategy for each component.

Solid surveys

Fleet- or building-graphics surveys and initial sales interviews help gather critical information. In these initial meetings, the salesperson must learn the key players who influence a buying decision; what, if any, problems they're having with their current graphics vendor; and their decisionmaking structure.

For some programs, a graphics survey helps develop an account strategy. A survey can unearth the current program's graphics, such as fleet, wall, window, floor and POP applications. A survey can also identify strengths, weaknesses, inconsistencies and deficiencies in current programs. Subsequently, you can develop suggestions and designs for improvements.

Major changes (or problems) trigger key opportunities, such as a new logo or a new advertising program. When inspecting current graphics, examine the decals' conditions. Durability problems can result from wrong materials, improper substrate preparation, poor application techniques, incorrect cleaning methods or harsh environmental conditions.

Solutions to these problems are prime opportunities for sales. Photograph cracking vinyl, edge lifting, tenting around the rivet heads, fading and other problems. Your sales presentation should provide solutions — material recommendations, engineering and processing alternatives, or care and cleaning recommendations.

Removing old vinyl graphics and installing new decals can always be problematic. Faraway fleets and stores can further complicate the program. Learn where the vehicles and stores are located, and when they're available for installation. Discern what problems the customer has faced and procedures used in prior applications. Naturally, indoor facilities are advantageous.

Some programs require storing, packaging and shipping decals to various locations. Typically, the fabricator provides the customer with a monthly inventory report that lists beginning inventory, materials sent and ending balance.

Because programs requiring storage can be very complicated — and create unhappy customers should problems arise — ask detailed questions to learn and understand their expectations.

Appropriate questions include:
- Is the customer satisfied with the current program's management?
- What would they change about their program?
- What types of program complaints have they received from their other corporate locations?
- How quickly do they receive their graphics following a release or order?

When gathering information, understand prospective accounts' climates. Do they have a reason to change? You win graphics programs when existing programs have problems: product performance, customer service and program administration. When corporate graphics become outdated and require a facelift, design can be a selling tool. Sometimes, price alone can spur change. However, customers will rarely change suppliers to save a few pennies; besides, who needs that sort of client?

The willingness to change indicates dissatisfaction. Understand the problem, and then define remedies in your sales proposal. If a prospect enjoys all aspects of his present program, you have little opportunity for business.

The only way to alter the company's homeostasis is to demonstrate hidden problems or potential trouble. Your sales proposal should then solve these problems.

Persuading decisionmakers

Some compare the sales process to an attorney presenting a case before a jury. In the survey process, you collect evidence. The sales presentation involves making your case. Next, the deliberation phase begins, and, finally, the buyer — acting as the jury — weighs your proposal's merits, and gives the ruling, hopefully in your favor.

When developing contractual agreements, fully understand the customer's application, the project's durability requirements, the environmental demands the graphic will endure, and, in the case of digital printing, the level of photorealism and color reproduction the customer will accept. Equally important, the sales presentation must demonstrate to the customer that you can satisfy these needs and expectations.

Accurately convey the finished product(s) your company can deliver. Salespeople commonly oversell and over promise their company's capabilities, sometimes to the point that customers' expectations exceed what can be delivered by today's technology.

Remember, when you construct a sales order, you create a contract that binds you to a stated level of performance. When you overstate that performance level, and misrepresent the products and services to be delivered, the contract's terms are violated.

A salesperson brings value to the buying process by acting as the buyer's consultant. A salesperson's job is to make the buyer's decision easier by explaining their options and providing information required to make a decision.

When vying for a job, you'll often present your case at nearly the same time your competitors present theirs. Those evaluating your proposal will assess the differences. The buyer most likely has set decisionmaking criteria. Do you know the criteria, and does your proposal satisfy the prospect's standards?

I worked for one successful salesman that always included a "Criteria" section in each proposal. This section listed all considerations the prospect should ponder. By introducing a new set of standards, he changed the rules, thereby influencing the outcome. Subsequent proposal sections provided solutions and information that satisfied each criterion.

To make his case, he covered:

- Consistency and control: How will you ensure faithful reproduction of the client's corporate image, from vehicle to vehicle, store to store, year to year?
- Durability: Graphics must endure the program's entire lifespan.
- Machine-made vs. handmade: Make sure customers understand machine-made graphics are much less prone to mistakes.
- Safety: As an example, he cited that trucks identified with reflective, vinyl markings provide motorists with advance warning that saves lives.

Your sales proposal should emphasize what sets you, your company and your sales proposition apart. Why should the prospect select you over your competitor?

As you're concluding the deal, always make it easy for the prospect to make a big decision by first getting him to make decisions on a program's minor details, such as design selection, color choices, approval of full-size production art, quantity, delivery date, and an agreement on the terms or conditions.

Conclusion

As in all professions, technology is revolutionizing sales. Some believe that, as e-commerce develops, outside salespeople won't be needed. I don't believe that will happen in my lifetime; making a deal is a face-to-face activity. Graphics sales require attention to detail, establishing and maintaining personal relationships, and understanding dynamics within an account. Further, it's very difficult for a customer to order graphics over the Internet or phone.

Selling vinyl graphics requires an intimate understanding of the buyer's needs and problems, and the buying process within a particular organization.

Chapter 2
Prospecting

Companies try different ways to grow their business — some wait for prospects to walk through their doors, and others pursue would-be customers. It's much easier, and less stressful, when customers contact you.

But this places your company's destiny in the hands of fate. Being proactive makes your own luck. Finding new prospects requires more effort, but it's a faster way to grow revenue.

When I worked for one fleet-graphics company, every manager from the president down hit the phones for an hour or two each day. These prospecting efforts contributed to phenomenal growth — in the company's first three years of business, sales grew from $800,000 to $1.8 million to $3.8 million. Today, it's one of the largest fleet-graphics companies in the United States.

Diligent prospecting can yield the proverbial "golden egg" for your shop. However, if left unattended, your leads can easily be devoured by competitors.

Most salespeople know prospecting is effective, but few make it part of their daily routine. Most people dread picking up a phone and encountering frequent rejections. Because phone prospecting can be excruciating, salespeople invent every excuse not to do it. Here are few I've heard recently:

- "The leads are no good."
- "It's summer; everybody is on vacation, getting ready for their vacations or are just returning from vacation."
- "That's telemarketing; that's not my job."
- "I don't have the time to do it."
- "I have more important things to do."

What could be more important in sales than discovering new business opportunities, determining a prospect's needs or identifying a company's decisionmakers? Isn't that part of being a salesman? Of course it is. But, it's easier to stay in a comfort zone and call on familiar customers rather than to speak to strangers.

Some salespeople fear catching a prospect on a bad day and stoking his ire. Honestly, what's the worst thing someone could say? "We don't have any trucks"; "We're not interested"; or "We're happy with our current supplier." Granted, someone giving such a response probably won't yield a sale, but these words are hardly fatal.

Prospecting strategies

As with any program, establishing objectives is critical. Prospecting simply qualifies sales leads and separates those companies that might buy graphics from those that probably won't. That's it. The primary goal isn't to sell graphics over the phone, nor to set sales appointments. The key objectives of phone prospecting are to:

- Discover whether a prospect can use your product. Do they have a fleet of trucks, a chain of stores or other signage requirements?;
- Identify the key decisionmakers; and
- Maintain contact with prospects that might need future graphics.

The phone process, which has worked for me, involves qualifying leads over the phone, following up with an information package, then calling to ask for an appointment. This approach requires preparing a follow-up, direct-mail package, which should include a cover letter, before you start calling.

Sending information prior to asking for the appointment helps familiarize the prospect with your company and services. It also helps stimulate interest in your business, which facilitates setting more appointments and, subsequently, creating more opportunities to make presentations and proposals that create more sales.

A few days after sending your kit, call again to see if the prospect has received your mail and is interested. Any delay in your follow-up will result in the prospect losing interest.

Only schedule an appointment if there's real interest — this prevents wasted sales visits. Direct-mail programs don't automatically require phone prospecting; some salespeople successfully follow up with e-mails.

Like mineral prospectors, successful lead prospectors must use the proper tools, such as calling referrals from existing clients and networking within industry-related associations.

The numbers game

We've all heard that sales is a "numbers game." If you contact more prospects, you'll set more appointments and have more opportunities to bid and, consequently, make more sales. Today's popular sales literature declares quality calls are more important that quantity. Targeting better prospects is important to sales success, but more is still better.

If one salesperson makes 25 new contacts each week, and another only makes one, who has the better chance of success?

If phone prospecting is part of your marketing program, set goals. When I worked for one screenprinter, we entered a realistic lead goal into our database and documented calls made and direct-mail literature sent. Quantifying our activities measured our progress and kept the program on track.

As the business maxim says, "What gets measured, gets done."

Use sales-contact software to log information. At the very least, design an account-profile sheet that lists such pertinent information as the company name, key contacts, phone numbers, e-mail addresses, number of stores or fleet vehicles, competitors, sales potential, follow-up steps, and when and where to make them.

Developing your prospect list

If you want consistent sales, keep the proverbial "sales funnel" full with new sales leads. Today's wealth of resources should provide ample prospects. Here are a few possible sources:

- Manufacturers' directories, such as the Thomas Register;
- The Yellow Pages;
- Former customers;
- Networking;
- Referrals; and
- Industry-related associations.

Directories

I have used directories of all types as a lead source. Every state where I've lived has a manufacturers' directory. Other directories exist for grocery, drug or convenience stores, as well as foodservice-distribution and trucking companies.

Don't forget the Yellow Pages — if you've been selling graphics programs to beauty salons or heating and cooling contractors, find similar businesses.

Industry associations

Other good lead sources are industry associations. As a fleet-graphics salesman, I belonged to the Calumet Safety Council, which comprised fleet managers. I was like a kid in a candy shop.

Referrals

If you've enrolled in sales seminars or read sales books, you've learned that referrals can be a valuable sales tool. Any salesperson worth his salt knows that, but few do it. When was the last time you asked customers if they knew of other companies in their area that could use your services?

I worked with one salesman who constantly did this to drum up new business. He would say, "I really don't know this area very well. Do you know anybody else around here with a fleet of trucks that I should be calling on?" He would call the prospect and make sure to drop his customer's name as his icebreaker. Will this technique work every time? Probably not, but what does?

Former customers

It doesn't hurt to ask a former customer how you can get back in his good graces. What's the worst that can happen? Can he threaten to take his business elsewhere? No, he's already done that. It never hurts to give an account one last shot.

Networking

It pays to have friends in the industry. One of the best sources for leads, leasing and trailer salespeople, will know long before you will who's buying and leasing equipment. In one year, the leads that I received from one Chicago leasing company amounted to $300,000 in sales, with a gross profit margin of more than $100,000.

The phone script

I've had my share of phone-prospecting success. But, I admit I'm not a natural — that's why I write a script. When I first started prospecting, I mistakenly winged it on every call. Without my talking points outlined in front of me, I frequently became flustered as I fumbled for the right words.

Using a script helps the prospector ask the right questions and convey the right message. Does this entail reading a script verbatim? Certainly not; it's important to sound natural.

Important Sales-Prospecting Points

- From the top down, make prospecting a regular part of your shop's routine.
- Leave your comfort zone. You can't grow business relying exclusively on existing clients.
- Use prospecting calls to separate strong prospects from weak ones.
- Send a direct-mail information packet immediately after prospecting, and follow up the package with a phone call within four working days. Any lag time can cause prospects to lose interest.
- Some sales "experts" claim that making a few calls to "quality" leads is preferable to making frequent sales calls — don't believe it. At the end of the day, someone with more prospects has more opportunities to make money.
- What gets measured, gets done. Keep a record of prospective customers, their decision-makers, contact information, potential sales and the type of follow-up required.
- Widen your circle of influence. Join appropriate business associations, enlist clients for referrals, network with others in the industries you serve — and never be too proud to ask a customer what you can do to win back his business.

So, given what I know about the market today, I sometimes deviate from the script and, maybe, embellish just a little. But, if I digress and stray off target, as I occasionally do, my script helps me collect my thoughts.

To write a script that sounds natural, impart a conversational tone. If someone gives you a canned script, rewrite it to fit your personality and style. After you write the script, practice it repeatedly, out loud, until you're comfortable with your pitch. You may even want to rehearse the script with a friend or loved one. With any script, first identify yourself and your company: "Good morning, this is Joe Jones from New Vision Graphics."

Phone protocol

Everybody has his or her own delivery, and several techniques can qualify an account. When I probe for information, I remain low key. I say, "I need your help. I understand that you operate a fleet of trucks. Is that correct?"

In my experience, when you combine humility with questions, the other person is usually much more receptive. If you're too pushy, the other person's natural response is to push back or be uncooperative.

If the company doesn't operate a fleet of trucks or a chain of stores, politely end the call and move on to the next prospect. If the company can use your services, probe a little further. My second question: "My company designs and manufactures truck graphics. Who would I need to talk to about your graphics program?"

Suppose the receptionist has connected you to the fleet or marketing manager. Next, confirm he or she is the right person: "Mr. Smith, I understand you're the contact regarding your company's fleet-graphics program." After pausing for a couple seconds, if he doesn't respond, ask, "Is that correct?" If he isn't, ask who is.

Before launching into your sales spiel or barraging your prospect with questions, first ask if it's a good time to talk. If it isn't, ask for a convenient time to call back.

The initial prospecting call's objective is collecting information, not delivering a sales pitch. Remember the words of Dr. Stephen Covey, author of The Seven Habits of Highly Effective People: "Seek first to understand, then to be understood."

As with face-to-face selling, open-ended questions (which can't be answered with a yes or no) help the prospect open up. Usually, these questions start with who, what, why, when, where or how. Once the prospect starts to talk, carefully listen to what he or she says. Above all, don't interrupt until the prospect finishes talking.

Nothing sounds better to a prospect than his or her name. During the conversation, try to call the customer by name a couple times. This helps you take the chill off a cold call and build rapport. Just don't overdo it to the point of annoyance.

After beginning the conversation, be prepared for the prospect to ask you questions. Phone conversations often unfold differently than planned. To overcome any obstacles, "improvise, adapt and overcome," as the saying goes.

Selling yourself

In many cases, the prospect will ask questions about your company. Prepare a list of characteristics that distinguish your company from your competitors. This list should emphasize the features, benefits and advantages the client will gain by hiring you.

If the prospect asks questions you can't immediately answer or has some objections to doing business with your company, write these questions and concerns down before formulating appropriate responses.

Craft your selling proposition around your business' uniqueness. For example, if you're an award-winning designer, emphasize how quality artwork can elevate your customer's image. If you've recently

The initial prospecting call's objective is collecting information, not delivering a sales pitch.

purchased state-of-the-art equipment, tout your efficient turnaround and product quality.

Whatever your message, keep it positive. Take the high road and make your customer feel positive; don't mudsling against your competitors.

As you finish your phone call, thank the prospect for his or her time. And, if it's a bona fide prospect, say you'll follow up by sending company information, and a salesman will call in a couple weeks. Log any qualified leads into your target-account list.

Leaving voicemails

When making business-to-business prospecting calls, the caller typically must leave a voicemail rather than connect directly to a live prospect. Personally, I hate leaving voicemails. But, when making a business call, it's often the only way to communicate with the prospect.

To avoid flubbing my words or babbling incoherently, I prepare voicemail messages in advance. In my experience, a prepared statement usually sounds much better than off-the-cuff rambling.

Conclusion

I've shared some of my successful phone-prospecting techniques. They've worked for me, but they're not the only approaches. So, if you have friends in the industry who've successfully drummed up new business, ask for their advice. Take them out to lunch to learn about their techniques. By getting your friend away from the shop, you'll avoid the inevitable interruptions.

Whatever approach you use, consistently make phone prospecting part of your daily routine. Regular prospecting will provide a few golden opportunities. The new accounts gained will help offset any losses from your current customer base. Regular phone calls to existing customers should complement your prospecting efforts. This helps generate additional sales and minimize account attrition.

Chapter 3
Fleet Graphics

Mention the terms "fleet graphics" or "truck graphics" and what usually comes to mind are the likes of UPS, Coca-Cola or Frito Lay – high profile companies with hundreds of trucks on the road. Sign-related business in connection with these large fleets is tremendous: so tremendous, in fact, that, overall, truck graphics consume as much pressure-sensitive vinyl as all other vinyl sign applications combined.

According to industry statistics, although large fleet with more than 25 units account for approximately three million commercial vehicles in North America, common carriers represent only one-third of that number. The rest are operated by private-industry, leasing companies, utilities, school systems and government.

Although the large fleets attract (or should I say "distract") the attention of graphics sales people, smaller fleets with fewer than 25 units make up a far larger market segment, according to the Transportation Technical Service, Inc. in Fredericksburg, VA. In North America, more than 10 million vehicles fall into this category. Locating these small fleets can be difficult, but the competition in this market segment is less severe – and the profitability therefore higher, with margins of 30 to 40 percent being easily attained for these smaller fleets.

Until the advent of large-format digital printing and vinyl plotters, the production of fleet graphics was the sole domain of a few dozen large screen print companies. Because one-time charges for color separations, dies, artwork, stencils, etc. made short runs impractical, what amounted to rolling billboards with four-color process pictorials could be economically produced only in larger quantities. For this reason, fleet graphics companies were inclined to target larger fleets at the expense of their smaller cousins.

Then all that changed. "Digital printing made the rolling billboard concept affordable for smaller fleets," says Larry Graefen, President of Michigan-based Harbor Graphics Corporation. Graefen suggests that one way for sign shop owners to take advantage of this untapped market without the required learning curve and a heavy investment of capital (not to mention the hiring of additional shop personnel) is by partnering with a whole-saler, who owns a large format digital printer. "A large format printer, such as an electro-static system, is a major investment. The training and expertise required to operate these

systems is also significant. It takes more to get into this business than writing a check and flipping a switch."

Another problem can be the actual locating of a small, private fleet customer. An effective way of unearthing potential new opportunities is to develop relationships with people in the truck sales and leasing business. "These individuals know before anybody else who is in the market for trucks," says Graefen. "Many leasing companies even include graphics in a lease package, thereby making otherwise unattainable graphics programs affordable."

Marketing strategies

Success in selling fleet graphics requires aggressive marketing. "A truck graphics opportunity rarely comes walking through your door," Graefen points out. "Many effective sales people develop a database of private fleets. Prospecting is often done on the phone or by direct mail."

After initiating an interest in fleet graphics, the actual selling of the project comes down to providing the prospect with the information necessary to make an intelligent buying decision. One of the chief reasons that fleet owners decorate their vehicle in the first place is that fleet graphics serve as an economical medium for effectively communicating a company's corporate image and advertising message. Additionally, vehicles that feature

Photo courtesy of Blue Media

reflective graphics can help to prevent nighttime accidents. Attractive fleet graphics can also have a positive impact on the attitude of drivers, employees and customers.

"Fleet graphics can play an important role in a company's marketing plan," says Tom Cornelius, National Sales Manager for FDC Graphic Films (South Bend, IN). "It can help a company increase brand awareness, and that in turn can help increase sales and profits." Fleet graphics achieve these results by virtue of the high visibility associated with road vehicles.

According to Cornelius, the actual number of visual impressions generated by fleet graphics depends on where the vehicles are operated. Tractor-trailer units driven on highways between metropolitan areas can generate 10 million visual impressions per unit per year. Delivery vans operated in urban environments such as Chicago or New York City can generate 16 million impressions annually."

"The 3M Company has sponsored several research projects on the impact of fleet graphics on the consumer," says Graefen. One study shows that 74 percent of those who see truck graphics develop an impression of a company and its products from the appearance of the company truck. The same study showed that 29 percent of viewers indicated that they would buy or not buy based on their impressions formed after viewing a commercial vehicle.

Because of the high visibility of trucks, vehicle graphics can be a very cost-effective marketing tool compared to other advertising media, such as radio, TV, newspapers and outdoor billboards. In advertising terms, a typical fleet graphics program can generate 1,000 visual impressions for less than a dime.

In addition to affecting buying behavior, fleet graphics can also reflect the pride of an owner in his company, as well as the attitudes of drivers and other employees. "When drivers have pride in their vehicles," says Graefen, "it pays off in better care of their equipment and safer driving habits.

Graefen believes that in maintaining a uniform vehicle design from truck to truck, year to year, vinyl fleet graphics provide fleet owners with several advantages over painted graphics. Since vinyl graphics are machine-made, lettering, colors and corporate symbols are precisely reproduced in high quality materials that usually last the life of a vehicle, resisting fading, street salts, washing chemicals, steam cleaning and abrasion along the way. Painted graphics, on the other hand, typically last between one and three years. Additionally, vinyl graphics can usually be completed in 25 percent less time than it takes to hand paint letters and symbols, and in the event of an accident, can easily be replaced.

Although every sign shop has access to the same raw materials and technology, what distinguishes one sign company from another is design. "The designer must clearly understand the customer's marketing goals," says FDC's Cornelius. "He then needs to translate those goals into a layout." According to Cornelius, the key elements that the designer works with are the company name, logo and brand names, slogan, trim and accent striping and pictorials. "It's important to keep the design uncluttered," continues Cornelius. "Slogans should be kept to fewer than six words."

Conclusion

Although uncovering fleet graphics opportunities and turning those prospects into cash takes persistent sales effort and considerable hard work, it can be nonetheless a rewarding business. The finishing touch to any fleet graphics program is professional vinyl application, which is covered in the "Vinyl Application & Removal" section of this book.

"Fleet graphics problems usually result from poor substrate preparation or poor application technique," Cornelius says. "Sign makers should take the time to develop the skills of decal application, or they should subcontract the work to a professional decal installer."

Even though pressure-sensitive films are very durable today, vinyl graphics will not withstand customer abuse. Cornelius believes that after the graphics application has been completed, sign makers need to instruct their customers in the proper care and cleaning of vehicle graphics. Strong solvents and excessively abrasive cleaners or high water temperatures/pressures can damage vinyl graphics. Moderation is the best bet in cleaning. Usually nothing more than a mild detergent and a soft brush or rag is needed to clean graphics. Hard bristle brushes should never be used. A rag moistened with mineral spirits or a household cleaner can be used to remove tar and oil before the graphic is washed and rinsed.

Chapter 4
Window Graphics

Photo courtesy of Avery Dennison

Twenty years ago, my sales manager advised me, "Go where they ain't." He meant that I should sell something that my competitors were overlooking. Of course, I didn't listen to a word he said. I rarely did. Instead of pursuing a lucrative niche business, I decided to fight over the unprofitable, low-hanging fruit with other salesmen who ignored their managers' advice.

However, someone in every group not only listens to good advice, but puts it into practice. A colleague decided to focus his sales efforts on window-graphics programs because everyone else focused elsewhere. Most "experts" agreed that window graphics were a waste of time and predicted that the salesman was destined for failure.

Color by Pergament (New York City) used Avery Dennison Graphics' (Hamilton, OH) MPI window film to produce window graphics for the 1,500-sq.-ft. storefront window of FAO Schwarz's New York City toy shop. They reverse-printed the hand-cut graphic, which allowed shoppers inside and outside the store to see the design.

Most professionals viewed window graphics as long-term projects that required considerable time. They entailed prospecting on the phone, conducting detailed sales interviews and site surveys, and planning the program's particulars. Who wanted to work that hard, especially when efforts pursuing such programs never pan out?

This salesman realized that, if window graphics sales were easy, everybody would have been promoting them. Through hard work and perseverance, his road less traveled eventually rewarded him with a customer base of many large, very profitable, national and regional programs. Window graphics can pay dividends for you, too, if you make the effort.

Prospecting

The first step in developing window-graphics clientele is prospecting. This involves working the phones to qualify the best program candidates and identifying the decisionmaker(s). If you're uncomfortable prospecting on the phone, consider using professional telemarketers. A professional can qualify hundreds of calls a week.

A friend recently hired three telemarketers to build a prospect database. The trio camped out at his business for a little more than a week. In that time, they qualified more than 2,000 companies.

They gathered names and e-mail addresses of key contacts. After they'd finished, he confirmed mailing addresses, phone numbers and fax numbers. He's now using this information for mass direct mailings, e-mailings and fax campaigns. The total cost to build this prospect database was less than $2,000.

Establishing contacts

Before you begin, you'll need a list of potential clients to call. Individual retail outlets or small store chains can become excellent prospects for signmakers. Your list of window-graphics candidates might include car dealerships, gas stations, beauty salons and similar businesses.

Window graphics are a challenging, yet highly profitable, niche of the vinyl-graphics market. Incorporating window graphics into bus wraps, such as this one that incorporates Avery's perforated window film, are an increasingly common application.

After developing a database, schedule direct mailings to stimulate interest in window and building-identification programs. These mailings should include a cover letter and a printed, four-color brochure featuring your work. Or, use mass e-mailings to deliver your message. This message should be concise and linked to your Website, if you have one. You can also attach a PDF document file to your e-mail.

The sales call

Whether using direct mail or e-mail, follow up with a phone call. Verify that they've received your mail, and, if there's an interest, schedule a sales appointment.

Signshops that pursue smaller retail chains will often find much less competition than when targeting high-profile clients. Such companies can provide excellent profit margins.

The initial sales call should reinforce the customer's interest in the program and help gather information. Developing a sales proposal and design concepts requires a site survey. A site survey supplies additional information and involves obtaining necessary measurements, taking photographs and completing a market analysis. In essence, you're conducting a "market analysis" by asking the right questions.

Some important questions include:

- What are the company's existing corporate colors, logos and design motifs? Are any changes planned in this area? What liberties can the designer take?
- What are the company's key advertising and marketing themes? Are changes planned?
- What are the store's key product lines? Which lines should be emphasized?
- What are the company's strengths and weaknesses? How is the business currently perceived by customers and employees? How would they like to be perceived in the future?
- What are the company's opportunities and plans for growth in the upcoming years?
- Who are the company's primary competitors?
- What challenges does the company face in the future from these competitors?

Your interviewee's answers should provide the information needed to execute a design program and reveal the retailer's long-term business and marketing objectives. More importantly, conducting the survey should provide insight into the retailer's personality. This should help you prepare the sales presentation.

By conducting a marketing analysis, you'll distinguish yourself from your competitor. For your customer, you should position yourself as a corporate-identity specialist, not the everyday signshop. By making this distinction, you can command higher prices. Remember, a specialist usually makes more money than a general practitioner even though both have an M.D. behind their names.

Sell the benefits

A complete store-identity package is often costly and difficult to sell; promoting such a project requires more "sizzle" in your sales approach. Rather than relating material and manufacturing details, build the sales story around benefits for the store owner. How can the graphics package satisfy the needs that were identified in the survey phase of the selling process?

The essence of the sales story — what the program can provide — is often the same for most stores. The program's primary benefit program is increased store traffic, sales and profits.

Any successful retailer knows that windows can be effectively used as marketing tools. Grocery stores sport handpainted, paper signs hawking weekly specials. Car dealerships' windows are frequently painted with bold, colorful graphics. Urban retailers of all types carefully arrange their products in showcase displays to entice bustling pedestrians into their stores.

Reinforcing identity

In the last 25 years, use of vinyl window graphics has grown steadily, replacing paper and paint. These window treatments are an especially effective part of corporate-identity systems. For instance, McDonald's® "golden arches" window treatments are a familiar and important part of its visual program.

These types of designs complement other sign systems, such as interior or wall graphics, building and canopy fascia graphics, awnings, fuel pump markings, and illuminated and non-illuminated signage.

Because window treatments offer the retailer a cost-effective approach to enhancing a store's appearance, they can often be tied in with the sale of other signage. For example, an illuminated sign could easily be combined with window graphics, wall graphics and aisle signage. In this way, a single sign evolves into a comprehensive program.

Photo courtesy of Avery Dennison

Businesses that want their image and message reinforced in traffic will opt for vehicle graphics that use windows for optimal visibility. Window graphics are an economical alternative to producing a comprehensive brand or store identity.

Vinyl's superior durability

Vinyl window graphics provide retailers with advantages over painted graphics and other traditional systems. Vinyl's durability is certainly much greater. Paint systems typically provide three years of durability; vinyl generally allows seven years.

Vinyl also offers a higher gloss level and retention, with much better, longer color consistency across multiple locations. Computer-aided design and manufacturing also ensures graphics integrity for images, logos and typefaces.

Creating a uniform look

Today's retailers face several marketing challenges, one of which is the customer's perception of the business. In an urban setting, a store competes with hundreds of others. Frequently, one business looks like another. Colorful, bold window graphics can help attract attention, even if a retailer has a less-than-prime location.

Moreover, a well-designed program can help a retail establishment project unique corporate identity at separate locations, even if they're dissimilar.

Good look, low cost

Compared to other sign systems, window treatments provide tremendous value at a very low cost. Material and labor charges are usually lower. Also, shipping computer-cut vinyl incurs lower costs than freight charges for a large wood, HDU or metal sign. Application and maintenance costs are also much lower.

Whether the graphics are installed on the window's interior or exterior, vinyl application is comparatively fast and easy. By contrast, conventional decoration methods often become major construction projects that can disrupt store traffic. Plus, unlike paint, vinyl graphics don't subject customers to noxious fumes.

Projecting the right image

Changes in public attitudes are very difficult and costly to measure. But, a well-designed identity package can give a store a facelift it needs to create a positive response. You don't need market research to know that consumers prefer shopping in a clean, modern-looking, cheerful environment. Employees also exhibit a more positive and productive attitude in a clean work environment. A building's graphics program, which includes window graphics, can contribute to changing the store's atmosphere.

Design considerations

Window-graphics packages should enhance a company's image and complement its atmosphere and decor. To achieve a retailer's marketing objectives, programs should incorporate or reinforce such existing design elements as corporate colors, logos, slogans and pictorials. Select durable, quality materials.

Technology has dramatically changed the sign industry's equipment and materials. However, basic design principles for effective store graphics have remained relatively constant. When decorating windows of a small, retail chain, you'll need to develop designs that accommodate multiple window sizes and configurations at various locations.

A workable design is often divided into several smaller sections, or modules.

Manufacturing, installation and aesthetic considerations commonly dictate the proportions of different sections. Typical modules might include various striping sections, the company logo, company name, product or service offerings and slogans, store hours and a pictorial. Often, the designer's layout of these sections fits within the narrowest window. Narrow designs, however, aren't always necessary.

Prominently placed, well designed window graphics — such as this one promoting this year's All-Star Game, held in Milwaukee's sleek, new Miller Park — can convey a message with great impact at a fraction of the cost of comparable promotional campaigns. Note that the mullions that separate the panes of glass don't diminish the message. Holzhauer Sign Inc. (Milwaukee) fabricated the graphics.

Large pictorials, for example, might cover huge expanses of glass that comprise several windowpanes. Metal mullions — vertical bars that divide glass panes — aren't usually decorated and are unlikely to detract from the graphic's overall visual impact.

If designs include copy, keep messages short and simple, with big, bold lettering. Copy should be easily read from a typical viewing distance. Remember that, when using a block-style letter, the maximum viewing distance for 1-in.-tall characters is 25 ft. Small copy and fine detail are generally ineffective, especially if they're printed on perforated window sheeting.

True colors

With window applications, light colors are more visible than dark hues. Don't apply large expanses of dark colors to windows, unless they're bordered by white. As an alternative to a white border, an artist can break up a large, dark mass with lighter colors.

Dark colors can absorb the sun's heat, causing vinyl-covered glass to rapidly expand while cooler areas stay rigid. With part of the window expanding and other areas contracting, glass breakage is possible. Large areas of dark color next to regions of very light, reflective tones can also cause heat gain and glass expansion. Films with extreme gray-scale differences can result in glass fatigue and early failures. I recommend breaking up a design with different hues rather than using severe contrasts of light and dark.

Reflective sheeting applied to glass also poses a breakage hazard. The rigidity of stiff, reflective material may not allow the applied area to expand at the same rate as uncovered areas.

Design formats

Window-graphic layouts commonly use four basic design formats. One format creates a valance comprising stripes and graphics panels along the top of the windows. Popular for strip-mall stores because of its high visibility, this format focuses on the top of the window as the prime design area — the lower part of the window is generally blocked by parked cars.

Another benefit is that striping in the upper portion of windows generally exceeds the reach of miscreants that might deface window treatments. Although retailers' fears of vandalism are usually groundless, and window-graphics vandalism is extremely rare, many retailers insist that graphics be applied inside store windows.

Another popular design format, called the "fourth selling wall," positions graphics along the bottom of windows to conceal racks or gondolas. These graphics create an additional selling wall while covering up an eyesore.

Third, business owners can feature showcase window graphics. Years ago, and less commonly now, graphic artists designed store windows as enclosed show-cases to display a retailer's merchandise. In window graphics, "showcase" refers to striping or graphics around the window's perimeter. As with showcases of yore, the vinyl showcase focuses on a product or group of products.

The final option — the pictorial format — memorably conveys the retailer's message. Compared to the words "computer store," a pictorial of a person using a computer better conveys the message that the store sells computers.

Large corporations such as McDonald's® — who have such a readily recognizable logo — can easily promote visibility with window graphics that contain minimal copy. Interior-graphic installations may offer greater protection against graffiti and vandalism, but the superior appearance of an outside application far outweighs the risk.

Inside or out?

Although many sign professionals disagree, window graphics are most effective outside. While interior applications offer protection from vandals, exterior placement offers more advantages than risks. Reflective glare or tinting diminishes the graphic impact of an interior installation. Applied outside, graphics simply look better. Look at gas-station window graphics, where installations are almost always on the exterior.

Interior-window graphics also require moving sales racks and merchandise prior to application, which slows installation and disrupts store traffic. Exterior applications, however, pose no inconvenience to business operation, and installers may apply them after hours.

Keep it simple

When presenting features, benefits and advantages of a graphic program, don't, in most cases, explain manufacturing details — unless the prospect asks a specific question. Store owners don't need to know that the graphics are made with pressure-sensitive, cast vinyl, and cut on a computer-controlled plotter.

When viewing the design, retailers focus on how the graphics will enhance their stores. They'll imagine customers crowding the aisles and clerks busily ringing up sales. They likely will not be thinking about vinyl and paint, unless you divert their focus with details of product specifications, manufacturing techniques and other technical data.

Present yourself as a corporate-identity specialist. Sell an image rather than simply offering stick-on letters and stripes that store owners can apply themselves. Window graphics are not a do-it-yourself hobby. The program should be positioned as a turnkey construction project no amateur should attempt. A professional graphics program deserves a professional installation.

Finally, when discussing pricing, never break out individual component costs. This makes it easy for prospects to shop and compare. Remember, you're selling a retail-identity program that complements existing marketing themes and dramatically remodels the business' appearance. Selling a corporate image should be more rewarding than peddling easily duplicated computer-cut vinyl.

Material selection

The variety, quality and performance characteristics of pressure-sensitive films available to sign- makers and screenprinters have dramatically changed over time. Today's window-

Photo courtesy of Brixen & Sons (Tustin, CA)

When discussing window graphics with a prospective client, keep the focus on how they will bring attention to the store and, in turn, increase store traffic. Brixen & Sons (Tustin, CA) created this window graphic.

graphic films can be decorated in various ways: screenprinting, digital imaging or airbrushing. Some films feature special adhesive systems that allow installers to reposition the graphics without application fluid.

Films with removable adhesive systems are also available for short-term promotion. Static-cling vinyl and low-cost, ultra-removable polypropylene films provide other promotional alternatives and allow fast and easy removal.

Perforated window films
Perforated, window-marking films are becoming more widely used, not only to decorate storefront windows, but also for bus and building wraps, POP displays and exhibit graphics. Made with small holes, these films allow store patrons to see outside, while passers-by view the printed graphic.

A perfect choice for large-format graphics, perforated window films can be decorated using myriad printers, from grand-format inkjets to smaller, thermal-transfer printers, such as the Gerber EDGE™. For best results when using these films, create large designs with easy-to-read copy. The film's holes often lose small details.

Clear films
Colored translucent and transparent vinyls create a backlit effect in the evening, when the store's lights are on. Clear films, which can be screenprinted or digitally reverse-printed for interior applications, are often used for two-sided window decals, such as signs posting store hours. Frosted-glass vinyl simulates the elegance of etched glass without the cost or skill required for the real thing.

Specialty films
Metallized vinyls are also used for window decoration. These eye-catching films come in hundreds of colors and pattern combinations.

To extend the outdoor durability of these promotional-grade films, protect the graphic with a clearcoat, such as Frog Juice™. If you spray this clearcoat on an applied graphic, first mask the graphic and surrounding window surface with a premium-grade application tape. Using an X-acto® knife, cut the masking 1/4 in. outside the applied vinyl's edge. Then, remove the tape covering the vinyl and spray the film. The clearcoat provides UV protection and seals the graphic's edge.

Gold-leaf films, such as the Vinylefx™ hammered leaf, Florentine leaf and large engine turn patterns replicate gilding's appearance at a lower cost, and UV inhibitors have enhanced the durability of current goldleaf films. The variety of pressure-sensitive products offers signmakers a potent arsenal of materials.

Wet or dry?
Prior to applying window graphics, clean the glass. But don't use Windex™ for professional jobs. Its residual surfactant and silicone can contaminate the adhesive, preventing a solid bond. Some soaps and solvents can also leave a residue that causes a similar problem.

Glass cleaning should be a two-part process. First, wash the windows with detergent and warm water. After washing, rinse the surface with clean water and let the windows dry. Next, clean the glass with a mixture of water and isopropyl alcohol. Dry the surface using lint-free paper towels.

Dry application is the most effective way to install window graphics. For trouble-free applications, select the right film. Some films feature low-tack, repositionable adhesives that permit an installer to apply vinyl graphics to windows without application fluid.

Many signmakers refuse to attempt dry applications for fear of trapping small air bubbles. Squeegee marks on the adhesive side of the film are also common, but neither is a problem. Tiny air bubbles will breathe out of the film after a week of warm weather. Don't speed up the process by popping these bubbles with a pin. The light shining through the pinholes will be visible from inside the store. Let time and temperature fix the problem. Squeegee marks also vanish once the adhesive has time to flow out.

To prevent bubbles, use professional-grade application tools, vinyl films with repositionable adhesives and proper installation techniques. A few tips:

- Always sharpen the squeegee's edge before beginning the application, making sure the edge is free of nicks. Small nicks generate small bubbles.
- When masking, be careful. Wrinkles and bubbles in the application tape create wrinkles and bubbles in the applied graphic.
- When squeegeeing, use firm pressure and overlap your strokes. When removing application tape, put it 180° against itself, taking care not to pull the vinyl from the window surface.

Certain circumstances require wet applications. Aggressive adhesive systems make vinyl installation difficult, if not impossible. Hot weather also complicates matters.

If a window approaches 100°F, the adhesive becomes unmanageable and, thus, repositioning is impossible. Misting the surface with clean water cools the surface. As the water evaporates, the temperature can drop 20° to 40°F.

If you must do a wet application, use a commercial application fluid instead of a home-made concoction. Despite hearsay, Windex is not an acceptable application fluid.

Riding the edge

When installing window graphics, don't apply vinyl over rubber gaskets or on the window frame. Film applied to these areas usually falls off. Allow at least a 1/8-in. space between the graphics and frame. Most window graphics shouldn't be edge-sealed (perforated window-marking films are an exception). Some manufacturers, such as Avery, recommend protecting these graphics with an overlaminate.

To complete the installation, seal the film's edges by painting on the sealer with a fine-tip narrow brush, such as a #2 lettering quill. Using an overlaminate and edge sealer prevents water from collecting in perforations, which can cause edge lifting.

Removal

Removing window graphics doesn't pose the challenges of stripping other types of graphics. A razor scraper and plenty of elbow grease should suffice. Chemical film and adhesive removers are also available for de-identification. A strong solvent, such as xylene, removes edge sealant. When using chemicals, mask off the window's edges to prevent any damage to window gaskets.

Some vinyl manufacturers recommend using a heat gun or propane torch to remove film. If you opt to do so, use extreme caution because overheating can easily crack windows. When applying heat to glass (especially when the weather is cool), you have only one question to ask yourself: Do I feel lucky today?

Conclusion

Visual impressions mold our perceptions of products and companies. More importantly, these impressions influence our buying patterns. As such, planned visual communications are retail necessities.

Bold, colorful window treatments help retailers create visual images. With a well-planned, window-graphics program, retailers can expect increased store traffic and higher sales and profits. For signmakers, the effectiveness of window graphics provides tremendous opportunities that shouldn't be ignored.

Chapter 5
Floor Graphics

Photo courtesy of Gallup & Gallup Floormedia (Toronto)

Walk into a grocery store and try not to notice floor graphics conveying an advertising message. This ground-level, POP signage has gained popularity because it stimulates sales. That's why so many grocery and discount stores use floor graphics to announce their hottest specials and introduce new products. Unheard of a few years ago, sales for this new medium are projected to reach $2 billion within five years.

Floor graphics boost sales by as much as 30%. Gallop & Gallop Floormedia (Toronto) fabricated these graphics using white FLEXcon material and clear overlaminate.

Why they work

Consumers notice the graphics, read their messages and respond at the cash register. Floor graphics are as effective as more conventional POP displays in stimulating sales of impulse products, such as snack foods, by as much as 10%. Sales for some consumer products are reportedly 20-30% higher at stores with floor graphics.

Some of the success can be attributed to the medium's novelty. Still, others believe that ground-level advertising is effective because most people naturally focus their eyes downward, looking where they walk as they push their shopping carts.

As an advertising medium, floor graphics complement and reinforce other advertising messages and campaigns. Compared to newspapers, outdoor advertising, radio and TV, floor graphics are cost-effective, with prices in the range of $12 per sq. ft. for the printed piece. By dovetailing with other marketing programs, floor graphics strengthen brand recognition. Plus, by reinforcing an advertising message, these graphics influence buying decisions — most consumers don't choose which brand to buy until after they enter a store.

Other options

Grocery stores aren't the only venue for floor graphics. Museums, shopping malls, nightclubs, airports and exhibit halls also use floor graphics to provide directions and promote products.

Sporting-event sponsors often use floor graphics as short-term signage, and, in warehouse or manufacturing environments, floor graphics can remind or warn workers of possible hazards.

With the floor-graphics market growing, a few companies specialize in this new form of advertising. Although large screenprinting shops produce many such programs, signshop owners aren't excluded from this emerging market. Digital printing is ideal for short-run graphics for promotions or clearance sales.

When designing an identification program for small retail businesses, include floor graphics in the store's décor package. They should be part of your standard sign program, along with window graphics, wall graphics, fleet identification and aisle signage.

Floor Graphics **31**

Floor graphics 101

Floor graphics are ideal for almost any store with smooth, nonporous flooring. Acceptable flooring surfaces include vinyl and ceramic tile, sealed concrete and finished hardwood floors.

When designing floor graphics, use big, bold and colorful graphics that will grab shoppers' attention. Photos and 3-D effects successfully generate visual impact and buying interest, and die-cut graphics are usually more interesting — and effective — than square-cut patches. Lettering should be bold enough for consumers to quickly read, with simple layouts featuring minimal words to communicate messages most effectively. If possible, white and light colors should be avoided in the design because scuff marks and grime are visible on light backgrounds.

The right stuff

Companies such as Avery Dennison and 3M™ offer a wide range of materials for floor graphics. These materials typically entail a vinyl print-media layer and slip-resistant over-laminate that must meet demanding requirements. Various floor-graphics materials are designed for standard, digital-printing technologies.

Because store personnel often install the applications, select a user-friendly adhesive system.

Ideally, the adhesive should reposition so the graphic can be easily installed at a wide range of temperatures. However, it must have a strong bond that won't peel off the floor. When removed, the marking should detach with little or no adhesive residue.

Photo courtesy of Gallup & Gallup Floormedia

Like screenprinted graphics, these substrates are often affected by temperature and humidity. Ideal shop conditions are ambient temperatures of 65-75° F and 45-60% humidity. When processing, the print media should be removed from its packaging, sheeted and allowed to relax at least 24 hours prior to printing. The conditioning period also allows the liner to stabilize by gaining or losing moisture.

Overlaminates

Floor graphics must withstand the extraordinary abuse of foot traffic, dirt, grease, grit and chemical cleaners. Several overlaminates are specially designed for floor-graphics applications.

To ensure a floor graphic is slip-resistant, overlaminates are rigorously tested. The standard industry test, the American Standard for Testing Materials (ASTM) D2047 exam,

Gallop & Gallop fabricated these floor graphics for more than 1,100 convenience stores. Floor-graphic films undergo the ASTM D2047 test, which measures the traction the media provide, assuring a non-slippery surface.

checks the coating's coefficient of friction. In layman's terms, this is the force required to move one material over another. In short, the ASTM test measures the traction pedestrians could expect when walking on a floor graphic.

Polycarbonate and vinyl — most of which are pressure-sensitive — are often used as overlaminate films for floor graphics. Printed graphics should only be laminated after the inks are thoroughly dry. In high humidity, extra curing time may be required.

Heat-activated overlaminates are also available. Clear protective films can be applied to encapsulate the front and back of the paper print. The overlaminates overlap the print on all sides by 1/4 in. or more to form a waterproof barrier.

Polycarbonate films without an adhesive coating are also used to produce floor graphics. A mirrored or reverse image can be printed on the film's second surface. After the ink is completely dry, a mounting adhesive is laminated to the print. In this construction, the film serves as both the print and overlaminate.

Applications

Installing floor graphics is similar to installing other pressure-sensitive, vinyl material. Prior to application, inspect floors for broken or loose tiles and uneven surfaces. To ensure good adhesion, floors must be cleaned using a commercial floor cleaner, such as a citrus-based product. Grease and tar can be removed with rubbing alcohol. Before the solvent evaporates, wipe the floor dry with a clean, lint-free cloth to prevent residue that could affect adhesion.

Use common sense when positioning the graphic. Although today's overlaminates are durable, they won't withstand continuous forklift and heavy-equipment traffic. In addition, graphics shouldn't be installed in areas that may get wet, such as in an entranceway. While overlaminates must meet slip-resistance standards, wet graphics can become hazardously slick.

Floor-graphics applications should always be applied dry; never use application fluid. To prevent edge lifting, always squeegee the graphics' edges twice. Following the installation, wax the graphics and floor surface to seal the edges and prevent moisture from seeping between the graphic's layers and floor surface.

Chapter 6
Warranties

A signmaker recently asked how vinyl manufacturers establish their films' durability claims. According to him, the film makers allow themselves considerable "wiggle room."

He asked, "How can a film have a durability of five to seven years? It's either one or the other. And what does 'up to five years' mean? Is it a five-year vinyl, or isn't it?"

I believe vinyl manufacturers' durability claims are merely educated guesses as to product performance under typical conditions. These guesses are based on their industry experience and various weathering tests. As sophisticated as these tests are, they can't guarantee how a vinyl film will perform in the real world.

Shops that specialize in vinyl applications should be very careful with customer warranties and claims and not make promises they can't keep. Carefully discuss your vinyl's durability, and don't allow customers to pin you down for a warrantee or guarantee

Thus, when signmakers ask me about durability, I tell them it's important to know more about an application and geographic location. Even with this information, accurate durability claims can be difficult.

Vinyl makers can't predict the future either, because they have no idea how signmakers and their customers will use a particular product. Despite all the testing that manufacturers conduct, they can't test for every variable. For this reason, their warranties limit their financial exposure should problems arise.

The responsibility is yours

I'm not suggesting that vinyl manufacturers try to avoid their responsibilities. Rather, most film manufacturers supply replacement material even when claims aren't justified.

Although materials manufacturers are responsible for providing quality products, as the primary manufacturer, you're responsible for selecting and testing your sign materials and determining their suitability for a particular application.

Furthermore, the more you process the materials via printing and clearcoating, the less responsible the vinyl manufacturer becomes.

When numerous products are used together, determining the cause of problems becomes more difficult. In such cases, finger pointing among the raw-material companies usually increases.

Carefully tell your customers about your vinyl's durability. If a customer tries to pin you down for a warranty or guarantee, you could be entering dangerous territory.

A former boss gave me some valuable advice: "Drop the word 'guarantee' from your vocabulary. We don't guarantee anything. There are no guarantees in life. While we're at it, forget 'refund,' too. We don't give refunds."

Moreover, be careful with any customer warranties and claims. Don't make promises you can't keep. It's best to under promise and over deliver.

Early in my career, I made a huge mistake by writing a very detailed warranty into a sales contract. In that written warranty, I specified that the graphics wouldn't fade, chip, peel or crack for five years. Furthermore, I outlined the company's remedy in the "unlikely event of a failure," which involved removing the failing graphics, and remanufacturing and reapplying the new graphics.

Yep, I was young and dumb enough to put all that in writing. The ensuing learning experience was very painful. Unfortunately, the graphics didn't quite live up to the customer's expectations. In fact, it was a complete graphics meltdown. Within three months, the clearcoat and ink had worn off the vinyl. Poof! The printing had magically disappeared.

The company claimed nothing was wrong with the vinyl, or the way it was printed or applied. The fleet owner's washing system caused the problem. There was no product failure. However, the customer didn't buy the vinyl company's explanation, and the case went to court.

By writing the detailed warranty, I had entered the company into a binding contract. When the graphics didn't perform as advertised, the company was liable for a breach of contract. For that reason, the judgment favored the plaintiff. The customer received the remedy I outlined in the sales contract.

Thank God for attorneys. Yes, I really mean that. And, no, I haven't lost my mind. From working with attorneys, I've learned to appreciate what they can do for a business. If you have to put something in writing, a good attorney will make sure that all the "i"s are dotted and the "t"s are crossed.

Implied warranties

If you think you can skirt any express-warranty issues simply by not putting anything in writing, guess again. Ever hear of implied warranties?

Even though you don't provide a customer with a written warranty, one may be implied simply because you've made a sign for a specific application. As a sign professional, your job is to understand customers' applications and design a graphics solution that will satisfy their requirements.

The previously mentioned written warranty wasn't the only mistake I had made. My first mistake was not conducting a thorough job survey. I knew that the buses were washed daily. But I didn't take the time to inspect the washing system. If I had, I would've noticed the 8-ft.-tall, 6-ft.-diameter brushes. At that point, I might have envisioned the hollow nylon brushes abrading through the clearcoat and ink system. With such information, I could've engineered a better graphics solution. Instead of a clearcoat, the graphics should have contained an overlaminate.

Nearly all states have adopted many of the Uniform Commercial Code's provisions. Although the code's purpose was to create some national uniformity for commercial transactions, each state's statutes will likely differ. If you have questions, talk to an attorney.

If you're not a U.S. signmaker, you might assume that implied warranties are just part of the crazy American legal system. However, your country's commercial code may be similar. In fact, Europe has its own consumer-protection laws. It does my heart good to know that friends in other lands are trying to keep up with the American Joneses.

Customers have responsibilities

While many laws protect consumers, a few protect manufacturers, such as signmakers. After all, without some restraints, businesses in this country could never flourish. Our customers, for example, aren't free to misuse the signs and graphics we create for them.

By instructing customers about typical do's and don'ts, you can avoid product failures. If customers misuse a product, it's their problem. However, don't expect customers to admit their own negligence. Many unscrupulous customers will try to make their problems yours.

Years ago, I investigated a complaint involving vinyl lettering that had peeled from several large construction signs. Not surprisingly, the construction company was demanding new signage. The signmaker was, understandably, sick at heart, because he thought he'd have to perform costly rework.

We discovered, however, that the construction company had varnished all the signs, and the hot solvent in the varnish affected the vinyl's adhesion. Because the customer misused the product, he, not the signmaker, was responsible for the damage to the signs. As a signmaker, you're responsible for understanding your customers' needs. However, customers have responsibilities, too.

Whether you're a vinyl manufacturer or signmaker, you need to stand behind your products, within reason. Your customers should also understand the meaning of "within reason," and what is and isn't commercially acceptable regarding vinyl graphics. Tactfully explain to your customers that the signage won't last forever. The film you're using might be a five- to seven-year cast vinyl, but this doesn't mean it won't gradually fade and lose some of its gloss. It's vinyl; it's going to fade.

Furthermore, no vinyl film will perform the same way in every application. It's impossible. And it's unreasonable to expect a vinyl manufacturer or signmaker to produce products that will satisfy every end user's expectations.

If a sign is subjected to regular spillage of acids or caustics, its vinyl will become brittle and crack — I'll guarantee that. If a customer cleans his sign with harsh cleaning chemicals or cleaners with abrasives or scrubs the graphics with a hard bristle brush, I guarantee he'll damage the vinyl.

What is reasonable?

When customers ask about a product's durability, they're usually seeking assurance that the sign or graphics they're buying from you will perform in a reasonable manner for their particular application.

What exactly, though, does "reasonable performance" mean?

This depends on the application and your understanding of the customer's expectations. I once had a customer with a tanker fleet on the southwest side of Chicago. The company used a five- to seven-year cast vinyl for vehicle markings. Every year it replaced its graphics.

This wasn't a product failure. This company's tanker hauled acids and caustics. After a year of repeated spillage, the plasticizer would leach from the vinyl, and the graphics would crack. For this application, this was reasonable product performance, and the customer understood that. The company knew the marking system's limitations because they were explained upfront.

When a product failure occurs, it's best to remember that, usually, neither your supplier nor customer is trying cheat you. Rather, they just want a fair shake in resolving the problem.

Chapter 7
Estimating & Pricing

Photo courtesy of 3M

When I worked in estimating and production planning for a large screenprinter, we established our selling prices on a "cost-plus" basis. This meant that we totaled burdened labor — an hourly rate that includes employees' salary and benefits, shop over-

Jim offers formulae and suggestions for establishing competitive pricing and increasing profits on many types of vinyl-graphics jobs, such as this truck, which is decorated with 3M®'s 680 reflective and 180 non-reflective films.

head and other expenses. We then marked up burdened costs, defined as anywhere from 15% to 40%, taking our profit and the salesman's commission into account.

This system's key advantage is that it covers your costs. Using a sliding commission scale also encourages salespeople to sell at a higher price. Selling at 40% mark-up, a salesman would earn a 20% commission. However, a 15% mark-up drops the commission to 5%.

While pricing on a cost-plus basis usually ensures that you don't lose money, its major limitation is that, without consideration of prevailing market prices, you could leave a lot of money on the table.

Other methods

Pricing on a cost-plus basis isn't the only way to set prices. For instance, I know two brothers who own a successful commercial-graphics business on Chicago's South Side. They used another company's published prices to set their selling prices. They believed that, if their cash flow at the end of their fiscal year was bigger than at the beginning, and if their return on investment equaled or exceeded their profit goals, they were smiling. How can you argue with success?

Many shop owners use some variation of this approach when setting a selling price. Also, they often use the Sign Contractors Pricing Guide (*available from ST Media Group Intl., Cincinnati, OH, 800-925-1110*) to estimate competitive pricing. To track market prices, an owner often asks someone to call his competitors for quotes.

For quick quotes, one friend in the fleet-graphics business simply multiplied raw material costs and a labor cost estimate by five. His rationale was, if he operated at a 20% cost of sales, he wouldn't get hurt.

A few shops base their pricing on customer flow. When demand is high and manufacturing capacity low (i.e. fewer competitors and equipment), a shop can demand steadily increasing prices until business drops off.

A friend in St. Louis recently suggested that, if you're getting every job you quote, your prices are too low. He's satisfied if he loses 20% of the jobs he quotes. He believes this figure represents "price-first" shoppers, who usually are the most demanding, slowest-paying customers.

Shop and administrative costs

Many market-driven pricing methods, such as those I've described, make good business sense if you're covering your costs and making an acceptable profit. At the end of the day, however, you'll never know if you made or lost money, unless you know your costs.

Big businesses grow by understanding their costs and asking customers to pay their desired price. Keeping a handle on shop and administrative costs — also known as overhead — is essential for establishing a burdened shop rate or hourly labor rate.

Shop and administrative costs comprise everything that isn't charged directly to a job. In other words, it includes everything except the raw

Action Graphics & Signs (Columbia, IL) decorated this welcome sign for Valmeyer, IL, using Mactac's Mac-Mark 9800 film. Jane Kolmer, the company's president, observed, "I think you'll find the vinyl signshop alive and well. I've always admired the talent of vinyl shops for the unique, mechanical quality of the business. Not only do we design signs, but we also execute the mechanics of actually producing the product."

hourly labor cost, raw material costs and the other direct costs, such as special charges for dies, outside artwork, and printing and laminating costs. This includes your salary (you're paying yourself a salary, aren't you?), plus managerial salaries and clerical wages. Also, factor in utility costs, payments on loans, rent or mortgage payments, depreciation on capital equipment, and the cost of office and shop supplies that aren't used as raw materials (Table 1).

The sum of these costs is your monthly shop and administrative cost. Your accountant or bookkeeper can readily provide this information. These expenses will also appear on your company's income statement. Because costs and productivity change over time, tracking these costs is critical, especially if your business is growing rapidly. I know of one fast-growing company whose monthly overhead increased from $50,000 to $85,000 in three years.

Determining productivity

The next step is totaling your hourly employees' available monthly production time. If you keep accurate records, you should know your shop's productivity or the percentage of hours actually spent on saleable graphic programs. This doesn't include time spent on rework, which falls into the returns and allowances category.

If you're not keeping records of time charged to jobs, start. All the graphics companies that I've worked for have used a tracking system, including labor tickets, on which employees were required to fill out listed hours spent on a particular task. At the end of each week, I totaled the chargeable hours.

Maintaining accurate records is essential if you plan to track actual direct expenses against your estimate, following the job's completion. Labor and material tickets (listing actual mate-

Table 1: Shop and Administrative Expenses

Salaries, commissions, non-productive labor
Rent/mortgage payment
Utilities
Vehicle lease payments
Equipment lease payments
Computer hardware/software
Loan payments
Depreciation
Repairs/maintenance
Office supplies
Shop supplies
Phone expenses (including mobile)
Advertising
Sales-related expenses (e.g. travel, entertainment)
Bad debt
Taxes, fees
Shipping/freight
Insurance
Miscellaneous expenses

rial used on a job), the estimate, production order, copies of design art, vinyl swatches, color matches and copies of receipts for other direct expenses should be kept in a job folder.

Chargeable hours are used to calculate a burdened shop rate — $50 per hour would be a typical example — which is figured into the estimate. Chargeable hours can be totaled from labor tickets, or estimated using your shop's average productivity rate. As an example, suppose that your potential monthly labor totals 500 hours, and the average productivity is 60%. As such, your productive shop time is 300 hours.

Any raw labor cost not charged against a job should be charged to shop and administrative costs. For instance, the staff may use 40% of 500 potential labor hours per month — or 200 hours — sweeping the floors, maintaining equipment, answering the phones, etc. That raw labor cost is added to the overall shop and administrative expense, along with rent, lease payments, office supplies, etc.

Burdened rates

To calculate your hourly burdened shop rate, divide average monthly shop and administrative costs by the number of productive hours. For example, if monthly overhead is $12,000 with 300 chargeable hours, then the burdened rate is $40 per hour.

By estimating the number of hours required to produce the graphics, and multiplying by the hourly burdened rate, we arrive at the job's burdened labor cost (Table 2). When developing cost analyses, be sure to place estimate and actual figures side-by-side.

To calculate burdened material costs, add all raw materials that will be

Table 2: Total Burdened Labor Costs

Burdened Labor Costs	Actual
Design art (This is a shop/administrative cost if you aren't charging for design art.)	1 hour
Substrate preparation	½ hour
Computer cutting/digital printing	2 hours
Weeding	½ hour
Vinyl application	2 hours
Packaging/delivery	½ hour
Total production time	**6.5 hours**
Burdened labor cost	**$422.50**

(Multiply production time by the burdened rate [6.5 x $65])

charged to the job, including drop-offs and material required for proofs. Drop-offs, also known as overage, include any material that is purchased for a particular job, but not used.

For example, if you buy a 4 × 8-ft. sheet of sign substrate for a 4 × 6-ft. sign, the unused 2 × 4-ft. drop-off is still charged to the job, even though you may save it for future use.

When using past drop-offs on future jobs, the material cost should be added, even though you've paid for the product. If you don't factor in this cost, your estimates won't be consistent when the same sign is reordered in the future.

To account for normal production losses, raw materials should be multiplied by a waste factor of 10% or more. The waste factor for digital printing, for example, can range as high as 30% to 40%. If you screenprint, you must assess the job's difficulty. Printing a job with a halftone, for instance, can be very difficult, causing high material loss.

I also recommend burdening the material an additional 10% to cover your costs for carrying the job, and to account for delinquencies. To achieve a true 10%, divide by 0.9. Any other direct costs should also be burdened at the same rate (Table 3).

Table 3: Calculating Burdened Material Costs

Remember to:
1) Include all material charged to the job, including drop-offs (material).
2) Multiply total material by waste factor.
3) Burden the material costs by 10% (divide cost by 0.9).

Burdened Material Costs

12 2 x 3-ft. sign blanks @ $8.71 ea.	$104.52
24 sq. ft. of vinyl at $.85/sq.ft.	$20.40
24 sq. ft. of application tape at $.10/sq.ft.	$2.40
Subtotal	**$127.32**
Waste factor (Multiply $127.32 x 1.1 or other multiplier that accurately reflects your typical material loss.)	
Total burdened material cost (Add subtotal plus waste factor.)	**$140.05**
(Burden materials by at least 10%; divide by 0.9 to do this)	$155.61
Burdened labor costs (Table 1)	$422.50
Burdened material costs $140.05	
Total costs	**$562.55**
Selling price @ 40% markup ($562.55 ÷ 0.6)	**$937.58**
Price per sign ($937.58 ÷ 12)	**$78.13**

Profit

Many shop owners feel that they are their business. Consequently, they feel guilty earning a salary and tacking on a profit margin, thinking they're double dipping. The owner and his business are separate entities, so the owner needs to earn a salary as an employee of the company. And, your investors — even if you're the sole investor — deserve a return on their investment in the business. Profit also provides your company with the resources

Table 4: Profit Calculation

Desired Profit	Required Formula
40%	Divide by 0.6
35%	Divide by 0.65
30%	Divide by 0.7
25%	Divide by 0.75
20%	Divide by 0.8
15%	Divide by 0.85
10%	Divide by 0.9

needed to expand shop facilities and purchase new equipment.

Total burdened cost, however, isn't the same as the selling price — at least it shouldn't be. If you establish your pricing on a cost-plus basis, Table 4 will help you calculate your profit margin.

To some, Table 3's formulation doesn't make sense. But trust me, it works. To mark up a job by 20%, it might seem more logical to multiply its burdened cost by 1.2 than to divide by 0.8. The following example will hopefully make things more clear.

Multiplying a burdened cost of $80 by 1.2 yields a selling price of $96. Our profit of $16 ($96 - $80 = $16) is only 16.7%, not 20% of the price ($16 ÷ $96 = 0.167). By comparison, dividing an $80 burdened cost by 0.8 results in a price of $100 (80 ÷ 0.8 = 100), with a 20% profit. The formula will become even more clear if you must split the profit with a salesman, and his commission is larger than your profit.

Closing-out a job

Some shops conduct job closeouts to determine if they made or lost money. This process involves totaling actual material costs and job hours and any other direct expenses, and comparing it to the estimate. Indirect expenses, such as insurance, utilities, rent and any other expenses not immediately related to the job, also must be factored into a shop's profitability. If the job lost money, you need to determine if the reason was an estimating error, poor planning or inefficient production.

If the problem is employee substandard performance, don't compensate by adjusting an estimating standard. Not only does this mask inadequate performance, it could make you less competitive on future bids.

It isn't necessary to close-out every job. Where I worked, we reviewed five or six jobs each month. This was enough to uncover and correct problems. When you have a problem, don't assign blame — fix what's wrong. Closeouts conducted like criminal investigations demoralize employees.

A good profit margin helps make a good day. Many sign companies sell into new markets, such as school banners, to expand their customer base. Action Graphics & Signs fabricated these banners for local schools using 15-oz., fluorescent banner material from Mactac. The banner in the foreground also incorporates Coburn holographic film.

The last word

A Charlotte, NC-based sign professional challenges his managers to consider how cutting prices affects their business. If you cut prices by 5%, how much more will you have to sell to make the same profit as you would by not cutting the price? Cutting prices usually results in working harder to sell more for the same amount of profit. So much for making it up with volume.

If you discover that your prices are too high for your market, look for ways to cut your material costs or overhead, not your profit margin. Reduce your material costs through better purchasing practices, such as taking advantage of volume discounts. Creative production planning, such as nesting parts, can maximize material usage. Improved process control can reduce rework.

A former boss gave me the following tips on building a base of loyal customers:

- Concentrate more on creating value-added benefits, and less on matching a competitor's lowball price.
- Justify higher prices by demonstrating a higher level of personal service and attention to detail.
- Go where the competition isn't. Sell into new markets, such as window or store-interior graphics.
- Expand your base. Find new products to sell your current customers.
- Do something different. Provide your customers creative design solutions that satisfy their needs, rather than offering the type of signage that everyone else is selling.

Chapter 8
Time Management

Scheduling and job planning are usually routine tasks for most signmakers and screenprinters. But, because most shop professionals must attend to other daily details, even the best-laid plans can go astray. Consequently, promises to customers are broken; phone calls aren't returned; appointments are missed, and new sales opportunities are lost.

If you can't get your work done and never find planning time, this chapter is for you. I hope my suggestions will help you better manage your time and business.

Because I worked as a construction manager for a builder in the late '70s, I thought I knew a little something about scheduling and planning. However, as my career progressed, my workload increased.

At one fleet-graphics company, I was responsible for estimating, production planning, purchasing and scheduling installations. Because the job overwhelmed me, I overlooked certain things that needed to be done.

Realizing my dilemma and frustration, my employer at the time instructed me about a time-management system, which changed the way I organized my life and helped me improve my performance. Over the years, I've discovered that many signage and graphics professionals implement some variation of this system.

What's your plan?

The key to good time management is to implement a system that organizes all your jobs and tasks. Of course, to make the best use of your time, you should delegate authority and avoid distractions.

Many people use a system that organizes their tasks into a primary-project list and daily schedule. I use Microsoft Outlook to plan my activities. Then I print out the lists and organize them in a large binder. In the binder, I also keep a calendar and a phone list of business associates and customers. As the day progresses, I can pencil in changes to my schedule. If I travel, I take the binder with me.

My primary project list contains most of the big programs I'll be working on for an entire year. Although I won't work on most of these projects for several months, I keep them in one place to prevent losing a project. I prefer to organize projects in this manner, as opposed to keeping track of several hundred loose notes.

The primary project list can be organized into different categories, such as estimates, design, production and installation. On my primary project list, I also include personal-achievement goals pertaining to finance and education.

My primary project list contains more than things I need to do. In addition, the list includes information about a job's deadline and priority. For each job, I also keep an e-file, which includes a plan of action — who's doing what, and when.

Don't rely on your memory

A person who relies on his/her memory usually isn't reliable. Although I have a good memory, I've learned to write down everything I need to do. My daily schedule includes all the jobs I'll be working on that day, as well as any phone calls or appointments I need to make.

In the morning, when I plan my day, I review my primary-project list, my "to do" list for that date and my calendar. Some people prefer separate lists for all their personal and business activities. I think this is silly — keeping multiple lists can complicate your life. It's much simpler to keep all your daily activities on one sheet of paper.

Some people work very hard, but accomplish very little, because they primarily focus on the least rewarding efforts and activities for their business. This is the major reason to establish priorities.

Whether you use a computer or manual planning system, assign a deadline for each project, even if you haven't started working on it. Projects without deadlines usually remain on the back burner and never get done.

In addition to setting deadlines, I prioritize my projects as either "high," "low" or "normal." The five or 10 most important projects are classified as a "high" priority — the jobs that get my most immediate attention. If you try to work on too many projects, you can't focus. When you complete projects, cross them off your list and move onto the next most important jobs.

Keep it simple

Countless, high- and low-tech systems for scheduling and job planning are available. Your job is to find one that's suitable for your business. According to personal-growth trainers, if you do something for 28 days in a row, the practice becomes habitual.

Using Microsoft Outlook to plan my day allows me to schedule planning sessions at the beginning of each day. Usually, I only spend 15 to 30 minutes reviewing and updating my plans. Because I update my plans daily, and keep up with what I need to do, this is usually all the time I need. Besides, I don't want to spend more time planning than working.

Over the years, I've worked with some bright and talented people, who have spent most of their workdays sitting in their offices contemplating their navels. They plan for every possible contingency. But, in the end, they don't put any of their plans into action. These people are classic examples of paralysis by analysis.

Most vinyl-graphics projects are usually easily managed and don't require complex scheduling. For other types of work, especially complex projects, I've used such planning and scheduling systems as Gantt and Pert charts. However, there's no reason to complicate the planning process. Besides, if you get too tricky, you'll probably confuse your employees.

When to plan your day

It's up to you whether you plan in the morning or evening. Butch Anton of Superfrog Signs & Graphics (Moorhead, MN) schedules his jobs and plans his day in the morning. He spends 15 minutes with his assistant prioritizing his work, reviewing what's been done and what needs to be done for each job, and then creates a list of daily tasks. He feels he can satisfy customer commitments, because he completes important tasks first.

One of the most successful graphics salesmen, with whom I'd worked, planned every evening. The next morning, he hit the ground running. He retired from the industry a multi-millionaire. Who can argue with success?

He believed that planning in the evenings allowed him to get a better night's sleep. Otherwise, he would wake up in the middle of the night, in a panic, wondering what needed to be done the next day.

Most sign companies use work orders or job folders to track jobs. When I worked for fleet-graphics companies, job folders contained the sales order, estimate, production order, the design's installation layout, color swatches, records of the actual materials used and labor tickets. Some signmakers keep all this information on a work order. If a job requires a vinyl application, you should keep records of any installation expenses, such as travel costs.

If you maintain job folders, anyone who removes files from the filing cabinet should record them on a sign-out sheet, along with their name and the date. By using this system, one fleet-graphics company minimized the number of times files were misplaced or lost.

Don't get distracted

It's easy to get distracted. Snail mail, voice mail and e-mail can be big distractions. Before I leave the post office, I sort through my mail. After I quickly scan it, I decide what is, and isn't, important. At least 50% goes right in the trash. I try to reply quickly to voice-mail messages from customers and business associates. However, I delete most messages from sales people, unless I need a particular product or service.

As for e-mail, I only open messages from addresses I recognize. If a message pertains to a project I'm working on, I copy and paste it in my file for a particular task. I place e-mails from, or about, customers into my customer file.

I refer matters that don't pertain to my job to the appropriate people. According to one of my managers, "If I'm doing your job, I won't have time to do my job."

Business associates can also be distracting. When my plate is full, I don't have time for company gossip or bitch sessions. Rather, I need quiet time. I suggest closing your office door and telling your employees and/or co-workers that you need time to get your work done. Most of the time, if they think for themselves, your associates can answer questions and handle issues.

It's also important to know your limitations. It's very easy for signmakers to say "yes" to every customer request. However, when you over-schedule yourself, it's impossible to meet deadlines. You need to learn when to say "no."

Whether you work for someone else or you're the head honcho, you need to ask your employer or employee to outline his or her primary responsibilities. Without this under-standing, no one can be effective. As an employer, you may need to delegate more authority — pass jobs off to subordinates.

Don't depend on others

Although I believe that delegating is important, don't be surprised if your assigned tasks don't get done. As he was leaving office, Harry Truman said that Eisenhower was in for a real surprise. He predicted that Ike would sit at his big desk in the Oval Office and make many pronouncements, only to discover that nothing got done.

I'm not suggesting that you micromanage your employees. However, you need to monitor their performance. I worked with one woman in purchasing, who kept the top of her desk perfectly clean. I have nothing against cleanliness. It certainly beats sloppiness. But I assumed she was doing her work. Instead, she was stuffing the paperwork that she didn't do during the day — invoices from vendors — into the biggest desk drawer at the end of the day. When these bills didn't get paid, our vinyl film supplier cut us off. For a graphics company, this is a tough situation. This was also an embarrassing situation for her boss…but I survived.

To make sure assignments were completed, a former boss kept files on each employee. Regularly, he reviewed each person's file with that person. If a project wasn't complete, he would ask when the job would be done.

Section 2
Materials

Chapter 9
Pressure-Sensitive Adhesives

Early in my career, listening to technical presentations about pressure-sensitive materials was excruciating. The jargon — terms such as crosslinking, dyne levels, peel, quick stick and wet out — was totally foreign. Sometimes, I must have resembled a wide-eyed, yet dumbfounded, cocker spaniel, with my head listing to one side as I attentively heard the words, but didn't quite grasp their meaning.

Understanding pressure-sensitive materials can be challenging, even for those with considerable experience. Pressure-sensitive constructions combine various components: adhesives, release liners, facestocks and carriers. The extensive range of adhesive-coated films, foils, papers and foam products, sold within the graphic-arts industry, includes cast and calendered vinyl, metallized specialty films, unsupported transfer adhesives, overlaminates, premasks and application tapes, double-sided foam tapes and supported mounting tapes.

Adhesives used in pressure-sensitive materials, such as Avery Graphics' (Painesville, OH) vinyl pictured here, should be evaluated for tack and adhesion properties. Different adhesives bond to diverse substrates differently; to form a strong bond, an adhesive must readily flow over, or "wet out," the substrate's microscopic surface pores.

Few people understand the basic principles of pressure-sensitive materials because few manufacturers have formal training programs. And many existing programs aren't very effective. This chapter will answer signmakers' questions about pressure-sensitive materials by outlining some basic concepts and translating some jargon. Hopefully, you'll baffle and impress your peers with newly acquired vocabulary and technical expertise.

All about adhesives

Simply put, an adhesive is any material that bonds two substrates. The egg yolk that fuses my plates together when I don't wash the breakfast dishes is an adhesive. The sap that oozes from a cut pine tree is a common example of a natural adhesive. Natural latex — the milky fluid that comes from plants, such as rubber trees — is used in making the adhesive in your shop's application tape. However, the scarcity of natural rubber during World War II gave rise to the creation of synthetic adhesives, including manufactured rubber and acrylic.

Synthetic adhesives are related to plastics. Plastics and rubber adhesives — both synthetic and natural — comprise gigantic molecules, called polymers. Polymers comprise thousands of small molecules, called monomers. The chemical reaction that combines monomers into a single macromolecule is called polymerization. The polymerized synthetic resin, used in the manufacture of adhesives, used in cast or calendared vinyl, for example, is an acrylic resin.

Macromolecule chains, formed in the polymerization process, can be joined together by bonds called crosslinks. The degree of cross-linking within an adhesive's polymer chains affects its internal, or cohesive, strength. Cohesive, of course, means that material sticks to itself.

A highly crosslinked adhesive typically has more memory. As such, if you apply pressure to an adhesive and stretch it in one direction, this type of adhesive has enough elasticity — or memory — to return to its original state. As crosslinking increases, the adhesive's viscosity increases. Viscosity is the friction (or resistance) that occurs when layers of fluid slide against each other.

Harder, viscous adhesives exhibit lower initial tack, making them ideal for hot-weather applications.

Weatherability tests, such as this performed by Avery's Molly Waters at the company's 14,000-sq.-ft. applications center, are essential for determining a film's adhesion properties.

Conversely, adhesives with less crosslinking are softer and flow more readily, with higher tack, but lower ultimate adhesion.

Stickier vinyl doesn't adhere as well as less tacky material. While this might sound like some Zen riddle from an episode of Kung Fu, it's one of those strange, but true, paradoxes of life. Tack and adhesion aren't the same.

The macromolecular structure also affects how the adhesive performs. Generally, adhesives with long molecules are harder, yet more pliable, and build quickly to maximum adhesion. It's difficult for these long, gangly molecules to arrange in an orderly pattern. Instead, they're like a tangled plate of spaghetti, which is good. The molecules' entanglement creates greater internal strength, compared to polymers arranged in an orderly, but stiff, crystalline structure.

High energy

Pressure-sensitive products work a little differently than other types of adhesives, such as contact cements, cold glues, hot-melt glues, epoxies and silicones. Diverse adhesives bond to substrates differently.

When I worked in the packaging industry, I dealt with cold glues that seal corrugated boxes or folding cartons. They formed a bond, in part, by being absorbed by the paper substrate. Other adhesive bonds form through a chemical reaction between the adhesive and adherent (the material to which it's trying to stick).

Pressure-sensitive adhesives bond by molecular attraction, then mechanically interlock with the substrate. During this process, the adhesive flows into the substrate's microscopic surface pores. To form a strong bond, the adhesive must readily flow over — or "wet out"

— the substrate's surface. The degree of wet out depends on the forces pulling in opposite directions.

One force is the natural molecular affinity between two different materials. The adhesive's internal forces fight this magnetic attraction. The substrate's surface energy determines how strongly it attracts the adhesive. High-surface-energy materials have microscopic pores with higher and more pronounced peaks and valleys than a low-energy material's smoother surface. The table below lists where some common signmaking substances fall within the energy spectrum (see Table 1).

Technical and product managers contrast a car's newly waxed finish with an unwaxed finish to describe how an adhesive wets-out on low-and high-energy surfaces. Low-energy surfaces are similar to a newly waxed car; just as water sprayed on the surface beads up on the wax, the adhesive beads out on the substrate. On an unwaxed vehicle, the water spreads out. Similarly, an adhesive readily wets out on a high-energy surface.

You can easily see wet out on an unwaxed car; however, an adhesive's microscopic wet out can't be seen. Time, temperature and pressure impact pressure-sensitive materials' wet-out properties. As the adhesive continues to flow naturally into the microscopic pores, ultimate adhesion increases. Rising temperatures accelerate this process; predictably, colder temperatures retard the process.

At room temperature, pressure-sensitive material starts to wet out after pressure is applied. Room-temperature application serves as the ideal environment. Sometimes, pressure-sensitive vinyl sticks without pressure, resulting in graphic deformation. When this happens, the term used (other than the litany of expletives that the frustrated and angry installer utters) is pre-adhesion.

Other times, pressure-sensitive vinyls won't stick no matter how much pressure you use, which occurs due to numerous reasons. For example, adhesion failure often occurs when you're trying to apply a graphic in cold weather. Mishaps also occur frequently when the surface isn't properly cleaned, or if the adhesive doesn't mesh with the application substrate. For example, acrylic adhesives don't wet out on untreated polypropylene and polyethylene.

Table 1: An Energy Comparison

Here's a list of selected high- and low-energy materials, with their corresponding dyne levels (numerical values that represent surface energy).

Low-energy materials

Substrate	Dyne Level
Polystyrene	36
Polyethylene	31
Polypropylene	29

High-energy materials

Substrate	Dyne Level
Stainless steel	700-1,100
Glass	250-500
Polyurethane paint	43
Polycarbonate	42
Rigid PVC	39
Acrylic	38

Stumped? Read the instructions.

To select the best pressure-sensitive product for a particular application, consider the environment confronting the tape or vinyl. Will it encounter UV light, extreme heat or cold, or exposure to water and chemicals? Application temperatures, durability requirements and cost are other factors to consider.

To avoid problems with vinyl or any other pressure-sensitive material, check its specs. The manufacturer's product-information bulletin, and application instructions should provide all the information you need, including application and service-temperature ranges, peel adhesion, cleaning recommendations and information regarding application surfaces. If you still have questions, call your distributor or the manufacturer.

Here's a list of some things that can affect adhesion:

- Substrate contamination. The surface must be clean and dry. If dirt, oxidation, plasticizers or release agents contaminate the surface, adhesion failure often follows.
- Surface texture. Bonding a sign or nameplate to a rough surface requires thick, double-sided foam tape or transfer adhesive.
- Substrate porosity. Porous sign substrates should be sealed. Otherwise, any moisture absorbed by the substrate could adversely affect the adhesive bond.
- Thermal expansion. Substrates with different rates of expansion can break the adhesive bond. One such example would be mounting an expanded PVC sign to a metal building fascia. The sign blank expands and contracts at a much greater rate than metal.

The pressure-sensitive sandwich

The pressure required to make a pressure-sensitive material work isn't its sole trait. Pressure sensitives are also distinct from other adhesive families because they comprise multi-layered constructions called "pressure-sensitive sandwiches."

Like a local deli that offers various sandwiches, the pressure-sensitive sandwich utilizes many different formulations. These constructions can be divided into two broad categories: self-wound and linered.

The simplest self-wound construction merely comprises an adhesive applied to a facestock, such as paper or a plastic film. This product category includes masking tape, paper and film application tapes, and self-wound overlaminates. There are two construction variations: priming the facestock's second surface — the adhesive side — and applying a release coat to the first surface.

Just as priming ensures a better final coat of paint, it seals the surface if you're coating a paper facestock. By sealing the paper, less adhesive soaks into the facestock; thus, more adhesive stays on the surface. This allows more consistent performance throughout the roll. Because the adhesive lays atop the paper, there's more adhesive, which improves the adhesive's cold flow.

Priming also enhances the adhesive's bond to the facestock. A better adhesive anchor means that, if adhesive touches adhesive, less adhesive delaminates when you pull them apart. Also, with better anchorage, less adhesive transfers to the substrate after the tape is removed.

Some self-wound constructions are release-coated to provide easier unwinding; also, the process prevents tape from "blocking" on the roll. Blocking occurs when a self-wound product sticks together and doesn't unwind. Self-wound products include masking tapes, application tapes and surface-protection films.

Linered materials include a siliconized release liner, such as cast or calendered vinyl, double-sided foam tapes, supported and unsupported transfer adhesives, and linered overlaminates. The release liner protects the adhesive from the pressure-sensitive material, stabilizes the construction during conversion operations and smoothes the adhesive to provide the clarity needed for overlaminating films.

Release liners are constructed from various base materials, such as polyester and polypropylene films, and densified or polycoated kraft papers. Like anything else, each material exhibits strengths and weaknesses.

Liners are manufactured in various thicknesses. Commonly used in the sign industry, 78-lb. liners have the flexibility necessary for plotter cutting, while 96-lb., polycoated liners are ideal for screenprinters. Heavier paper lays more flatly on the press, and the polycoating prevents moisture absorption. Thus, the liner doesn't grow and tamper with registration.

Like a deli with an array of sandwich options, numerous pressure-sensitive adhesives exist. Self-wound adhesives comprise an adhesive applied to a facestock (application and masking tapes are prime examples), while linered materials — such as double-sided foam tapes and cast or calendered vinyl — include a silicone release liner that separates the adhesive from pressure-sensitive material.

Among the linered, pressure-sensitive products on the market, the most basic include single-liner, unsupported transfer adhesives, or transfer tapes. A transfer tape is not the same as an application tape.

Transfer tapes merely entail an adhesive applied to a release liner, which some refer to as "glue on a roll." A fancier version is transfer tape with two liners, which requires users to remove one of the liners prior to application to a print, nameplate or graphic panel.

Transfer tapes are often used to manufacture polycarbonate control panels. These panels are printed sub-surface, which means the polycarbonate film is printed in reverse, on its second surface, or underside. Then, transfer tape is laminated to the printed side. The release liner protects the panel's adhesive until application. In this application, the two most commonly used adhesive thicknesses are 2-mil for smooth surfaces and 5-mil for rough applications.

A supported transfer adhesive takes its name from the film carrier sandwiched between two adhesive layers. In the digital-graphics industry, this product is usually called a mounting tape.

The carrier serves numerous functions. It stabilizes the tape during the lamination process, provides additional rigidity for the print and allows the use of two different adhesives.

Commonly, a mounting tape comprises the following layers: release liner, a removable or non-removable adhesive, film carrier, permanent adhesive and the second release liner. The permanent adhesive applies to the graphic, while the other side adheres to the substrate.

Dozens of double-sided, foam-tape constructions are designed for numerous applications. Typically, these tapes contain a foam carrier, coated on both sides with adhesive, and a release liner that protects the adhesive on at least one side. A foam tape with a "closed-cell" construction (which resembles enclosed bubbles) forms an excellent moisture barrier. In contrast, if the foam carrier has a sponge-like structure, which resembles burst bubbles, the construction is called open-cell. Predictably, this construction sucks up water like a sponge.

Pressure-sensitive materials carry either rubber or acrylic adhesives. Products used in the sign industry typically comprise solvent or emulsion-based systems. Rubber-based adhesives tend to bond more consistently to the substrates; however, acrylic-based materials are less susceptible to plasticizer migration, which can ultimately lead to adhesive failures.

The following types of foam carriers make product selection confusing: high-density polyethylene, neoprene, acrylic, urethane and vinyl. The key difference among them is the internal, or shear, strength of the material.

The last, but not least, type of linered products is paper and film products used in the sign, screenprinting and label industries. In addition to cast and calendered vinyl, several products use various facestocks, such as polyester, polypro-pylene and paper. The typical construction includes a siliconized release liner, adhesive, occasionally an additional liner, facestock and, sometimes, a top coating or corona treatment to enhance surface energy and improve ink stability.

Rubber or acrylic?

Would you like rubber or acrylic adhesive on your pressure-sensitive sandwich? Adhesive systems fall into these two broad categories, and each has unique characteristics.

Soft, rubber-based adhesives wet out well. Because soft adhesives flow readily, they coat the substrate and provide very high "quick stick" to a surface. Because rubber adhesives wet out better than acrylics, they require less pressure during application to form a bond. A rubber-based system's tack allows adhesion to more surfaces.

The adhesive's tackiness makes it an excellent choice for working with low-energy plastics. As an example, if a shop needs to make decals for a manufacturer of polypropylene outhouses (as I once did), a vinyl with a rubber-based adhesive will probably adhere well to this plastic substrate. By contrast, vinyl with an acrylic adhesive is likely to fall off unless the plastic has been either corona- or flame-treated.

Rubber-based adhesives also exhibit consistent bonding to the adherent, meaning, adhesion values don't gradually grow as they do with acrylic adhesives. This is why rubber-based adhesives are used when making premasks. The adhesive bond doesn't build on the vinyl graphics during long-term storage. Otherwise, you'd have difficulty removing the premask.

However, rubber-based adhesives are susceptible to plasticizer migration. Thus, if you've applied a vinyl graphic with a rubber-based adhesive to a vinyl banner or flexible-face material, the banner's plasticizers could soften the adhesive to the point of failure.

Also, rubber-based adhesives are vulnerable to UV degradation and oxidation. As oxidation and UV weathering progress, rubber polymers start to break down and, eventually, cause adhesive failure. Oxidation also causes rubber-based adhesives to yellow; that's why your application tape yellows when exposed to light. The yellowing caused by slight adhesive degradation usually isn't enough to be an issue.

In contrast to rubber, acrylic adhesives withstand higher temperatures, exhibit good shear and resist the degrading effects of UV light, plasticizers and chemicals. However, they don't stick well to low-energy surfaces.

Although acrylic adhesives cost more than their rubber counterparts, they offer a much wider range of performance characteristics. Because signshop owners and screenprinters have diverse needs and applications, vinyl manufacturers provide their customers with an extensive product menu, including vinyl films with permanent, removable and repositionable adhesives.

Vinyl makers blend their adhesives and provide end-users with products offering specific attributes. For example, highly aggressive, permanent adhesives can be coated on brittle, ultra-destructible films to produce safety labels. This construction is so fragile that it breaks down into tiny pieces when someone attempts to remove the label.

Repositionable vinyl films make graphics application significantly easier, because the installer can snap a material back onto the substrate and then reinstall the material without damaging the graphic. The latest improvement features an adhesive with a microstructure of tiny tunnels. These tunnels make graphics applications easier because they provide channels through which air can escape. The result is a bubble-free, vinyl application.

Coating technology

Rubber and acrylic adhesives are available in several varieties. However, in the sign industry, the only widely used types are solvent and emulsion.

When adhesive components combine with a solvent, the solvent dissolves the mixture and evenly dispenses elements to form a solution. Solvent-based systems have been the mainstay of the pressure-sensitive industry for decades. Today, manufacturers must deal with solvent recovery and disposal.

By comparison, water-based adhesive systems are emulsions. Fine particles of the system's solid components are suspended in water. Rubber-based adhesives that are coated on paper application tapes are also emulsions. Further, acrylic emulsions, which have been used for years, are becoming more popular because they're more cost-effective and contain no VOCs.

Product testing

Product evaluation includes tests ranging from visual inspection to precise measurements. For example, vinyl and overlaminate manufacturers inspect adhesives for clarity and color.

When performing a light-transmission analysis, the measured light transmitted through the test sample is compared with a reference sample. An adhesive's clarity is important for such applications as window graphics.

Technical managers evaluate an adhesive's physical properties using "destructive" tests. The two most common tests used in evaluating a cast or calendared vinyl check "tack" and "peel." The Pressure Sensitive Tape Council (PSTC) established the guidelines for these tests, which are called PSTC-1 and PSTC-5.

PSTC-1 tests ultimate adhesion. In this exam, a strip of cast or calendered vinyl is applied to a stainless-steel substrate. An edge of the sample strip is pulled entirely against itself. The testing equipment measures the amount of force required to peel the tape from the panel. To test the adhesion growth over time, manufacturers check samples at other time intervals, such as 24 hours, 72 hours, seven days or two weeks.

PSTC-5 tests an adhesive's loop tack. During this trial, the two ends of a sample strip are placed in the jaws of the test equipment to form a material loop with the sticky side exposed. The test machine touches the loop of material against a stainless-steel plate and measures the amount of force required to pull the two apart. This tack test demonstrates an adhesive's aggressiveness. For vinyl manufacturers, this test gives technical experts an idea of how repositionable vinyl markings will be during application.

The rolling-ball test examines adhesive tack. A ball is rolled down the groove in a metal piece — shaped somewhat like a playground slide — onto a test sample at the end of the metal slide. A technician then measures how far the ball rolls on the sample pieces. The ball rolls farther on a harder adhesive than on one that's soft and tacky.

Other specialized test equipment measures the force (the release value) required to peel a pressure-sensitive film from the release liner. In developing a product, a manufacturer tests values at different times to verify any changes. High release values reveal difficulty when transferring a vinyl graphic from the liner, which can be problematic for the user.

At the other end of the spectrum, low release values indicate poor stability of the film on the liner. In this case, the vinyl can slip on the liner during plotter cutting.

In many cases, test samples will be tested in their "natural" state, compared to when they've been "aged." The natural state means the tape or vinyl-film sample comes right off the roll. In contrast, an "aged" sample may have been cooked in an oven at, for example, 120°F for two weeks.

By repeating the original tests after aging, examiners can predict how the product will perform after being on a distributor's shelf for a couple years. Experts look for drops in adhesion values or loop tack, an increase in release values and any changes in the facestock's physical properties.

With double-sided foam tapes, analysts conduct tests measuring shear, tensile and cleavage.

Shear is the internal strength of an adhesive or foam carrier. A shear test measures parallel forces generated within an adhesive or a foam tape. For example, a shear test calculates the force acting upon a tape holding the weight of a heavy sign against a wall. In testing the shear of a vinyl adhesive, a tape bearing a 500-1,000-gram weight is applied on top of the vinyl. If the adhesive doesn't hold the facestock to the substrate for the specified time, the technician must determine whether it's a cohesive or an adhesive failure.

While shear measures parallel forces, tensile measures the perpendicular forces imposed on an adhesive bond. Imagine grabbing two sides of a sign and pulling it away from the building. In this example, the stress is distributed equally over the entire area held by the adhesive or foam tape.

Now, instead of pulling on two sides of the sign, suppose you pull only at one edge. In this case, all your force is focused at one point. If you can break the bond at this point, you'll likely tear the sign off the wall as easily as you would unzip a zipper.

The strength of two materials bonded together at such an end point is called cleavage. I've torn apart rigid demonstration panels held together with high-bonding foam tape several times, not because I'm exceptionally strong, but because I understood the concept of cleavage.

Chapter 10
Embossed-Adhesive Films

Photo courtesy of Avery Dennison

A recent trucking study reported that more than 40% of American over-the-road trailers have no graphic identification. The president of a large fleet-graphics provider estimates that another 40% are poorly marked with minimal graphics. If you doubt these statistics, conduct your own survey on your next roadtrip.

How about Avery's EZ-Apply films for $800, Alex? Less than half of the film's surface area touches the substrate, making it easier to reapply than conventional films. EZ-Apply comprises microscopic tunnels that allow air to escape.

To an extent, tough economic times have relegated many graphics programs to the back burner. However, some vehicles represent difficult applications signmakers simply bypass. Labor costs can also make programs prohibitive.

Many signmakers avoid vehicles with rivets or corrugations like the plague. Regarding vehicle wraps, many say, "Forget it!" However, embossed-adhesive vinyl films could alleviate such fears and open up new fleet-graphics opportunities.

Products featuring this new adhesive technology include 3M's Controltac® films with Comply™ Performance, Avery Graphics' films with EZ-Apply™ Technology, Oracal's RapidAir line and MACtac's Bubble Free series. These fleet graphics vinyls feature adhesive systems with micro tunnels (also called air-egress channels) in the adhesives that help prevent air entrapment (otherwise known as bubbles) from forming, and make vinyl application faster and easier. With a little TLC, finished products are usually bubble- and wrinkle-free. If you make a mistake, forgiving adhesives let you reposition the graphic without distorting the vinyl facestock, which gives you another shot at success.

This chapter covers how these new adhesive systems work; how such higher-priced products ultimately reduce costs; and what you need to know when applying these films.

How it works

These products are easy to use because their adhesives are embossed with a network of microscopic tunnels. According to Tim Doyle, technical marketing manager for Avery Dennison's signage division (Painesville, OH), embossing the adhesive causes less than half its surface area to touch the application substrate, which cuts the adhesive's grabbing power accordingly. This means the installer can more easily reposition the graphic during the application process.

Tunnels facilitate application because they provide escape routes for air trapped underneath the vinyl. Air is expelled when the graphic is squeegeed.

How do they create those tiny tunnels in the adhesive? Remove the backing from the film, and look at the release liner's siliconized surface and adhesive. The liner paper is embossed with a network of minute grooves. The adhesive's surface, which resembles coarse fabric, carries the same impressions.

To create air tunnels, the release liner is first embossed with the pattern. Avery also prints blue dots on the liner. Next, adhesive coats the liner, which is cured and married with the vinyl facestock.

During this process, the adhesive assumes the grooves' pattern. The dots are also transferred to the

Do you need to balance work and play? With too much emphasis currently on the former? Try embossed-adhesive films, which allow users to easily reposition graphics, and thus reduce time and costs.

adhesive. The dots hold the adhesive away from the substrate, which prevents pre-adhesion until the installer applies squeegee pressure.

3M's films work similarly to Avery's blue-dot technology. Microscopic glass beads on the adhesive surface form a barrier between the adhesive and substrate. Squeegee pressure pushes glass-bead fragments into the adhesive, forcing it to touch the substrate.

These films aren't just for fleet-graphic applications. The technology opens doors for several large-format opportunities, such as wall, window and museum graphics. To satisfy various screenprinting, digital-imaging and signmaking requirements, the manufacturers have developed myriad products. These films are available with removable or permanent adhesives applied to cast, calendered or reflective films. Check 3M's and Avery's Websites — www.3m.com and www.averygraphics.com — to view their full product range.

Many new films accept screenprinting, electrostatic or solvent-based inks; top-coated films are available for waterbased printers. Because the pattern is embedded in the vinyl's first surface, thermal-transfer can't be used.

Printed graphics are perfect for full-vehicle wraps. In lieu of purchasing equipment, consider partnering with a large-format, digital-graphics provider. For example, one truck-graphics company owner outsourced large-format, digital jobs for years. After he built his business base, he could justify the equipment investment. Today, he owns two 3M Scotchprint® printers.

Heavier films with poly-coated liners meet the needs of fleet-graphics screenprinters. Such films as Avery's EZ 1000™ fleet-marking film are designed for easier application and removal, because removable adhesives simplify stripping old graphics. Some can even be removed from paint without chemicals.

Premium products = lower costs

Although these easily applied films carrier higher price tags, I can vouch they're worth every cent to you and your customers. Using them can cut application time and cost by half. That alone should be enough to sell your prospects.

When presenting the films' benefits, stress that faster application reduces equipment downtime. This means your customers' trailers are on the road generating revenue rather than tying up valuable shop space.

Stripping these graphics also takes less time and expense; they reportedly come off cleanly by just applying heat. With no adhesive residue, you won't need expensive, potentially dangerous chemicals.

According to Dave Harris, Avery's sales manager, 17% of McDonald's employee-installed graphics are trashed because of installation mishaps. Embossed-adhesive films won't turn burger flippers into professional decal installers, but when an amateur is given such graphics with a little training, the finished product can be quite attractive.

Training new shop employees could become easier. Theoretically, simplified application and removal could enable you to handle twice as many jobs. The productivity improvement inevitably will help your top and bottom lines.

The user-friendly adhesive system may encourage you to revisit such tough jobs as vehicle wraps. "Upsell" prospects by telling them that removable films protect original paint jobs. The paint doesn't lose its gloss; colors don't fade, and there are no "ghosts" from old markings. Trailers with new-looking paint jobs are easier to sell for a premium.

Vehicle wraps obviously cost more than applying a logo, basic lettering and accent striping. However, embossed adhesives help cut costs. If the graphic's cost impedes closing the sale, consider rolling the expense into the vehicle's leasing terms.

No bubbles

Before using any product, review manufacturer literature. Application instructions for their new products are available on 3M's and Avery's Websites. (Most application instructions cover substrate preparation, and application temperatures for films with traditional adhesive systems. These instructions also apply to the embossed adhesives.)

One key difference: Comply™ and EZ-Apply™ films bond to a substrate much more slowly than other films. The adhesive bond builds in stages and may require days — even months — to form an ultimate bond.

The slower bonding rate can be problematic when decorating some vehicle substrates. An additive to low-surface-energy clearcoats, on such vehicles as the GMC Sierra, can inhibit adhesion. Embossed systems will take significantly longer to form ultimate adhesion on such clearcoats than non-embossed adhesive systems.

Until ultimate adhesion is reached, the bond to these surfaces is vulnerable. To accelerate the process, especially in cold temperatures, try warming the graphics with a heat gun or propane torch. The added heat aids adhesive flow, which increases bonding.

Adhesion is a function of time, temperature and pressure. Even after the graphics have been installed, tiny air channels still exist. However, installers can release bubbles by simply applying thumb pressure.

Application pointers

When using these films, I suggest application tape with lower-than-normal tack. For cut graphics, a medium-tack tape is probably your best bet. For large, printed graphic panels, you may want a low-tack premask. Remember, embossed-adhesive films take time to bond to a substrate. High-tack tape can easily pull graphics from the application surface and, in some cases, tear the vinyl film.

To apply films with embossed adhesives, 3M has developed Power Grip™ tools. These include the company's Rapid Applicator; Magic Pad™ rivet applicator; multi-pin, rivet, air-release tool; and rivet brush. I've tried them all, and I especially like the rivet brush. It has a thicker handle and stiffer bristles, and its ergonomic design puts less stress on elbow tendons.

When using a rivet applicator to mold film to rivets, don't twist the tool, or the pad will twist right off. I destroyed a friend's Magic Pad in a few seconds of misuse. Just heat the film over rivet heads, and then press the pad over the rivet for a few seconds.

Personally, I use traditional application tools. But don't let my preferences dissuade you from trying these tools. They're inexpensive, and if they work for you, use them.

While Comply and EZ-Apply films are designed for use without application fluid, some signmakers who prefer wet applications must squeegee out all the moisture, or the fluid will be trapped in the air tunnels. Any residual moisture will freeze in wintertime temperatures, which could cause adhesion failure. In my opinion, there's no excuse for using application fluid.

The new technology isn't quite foolproof, but it's the next best thing. I conducted vinyl-application seminars using these films, and a few attendees managed to do what I thought was impossible. But two or three failures out of approximately 100 applications isn't all that bad.

Chapter 11
Vinyl Selection

I've never been a fan of calendered films. When I entered the industry, cast vinyl ruled the marketplace. Calendered film had a bad reputation — with good reason. These films shrunk so noticeably they left a tell-tale adhesive residue around an emblem — a "ring around the collar."

As older versions of calendered vinyls aged, they became brittle and cracked. Pigmented films didn't withstand the rigors of exterior exposure. They faded, chalked or changed colors.

In the past, one manufacturer's tomato-red, calendered film frequently turned dark burgundy when used outdoors. Previously, the stiff, non-conformable calendered

This wall-mounted directory sign was fabricated with calendered vinyl. It's advantages include higher tensile strength which provides greater tear resistance, lower cost, and easier application and removal.

films were much thicker. In addition to rigidity and color fading, your father's calendered films were prone to shrinkage.

However, calendered films have changed with the times. Today, they're thinner and more flexible. Some of the newer, glossier, calendered vinyls appear nearly identical to cast films. Plus, current calendered films shrink less and cause fewer edge-lifting problems.

Improved resins, plasticizers, pigments and additives have also improved calendered films' weatherability. All these improvements spur many signshops to use more calendered vinyl than cast.

However, appearances can be deceiving. While calendered films have evolved, some performance characteristics haven't improved. That's important to remember when your name's on the sign and your reputation's on the line.

By comparing the two major classifications of flexible, signmaking vinyls — cast and calendered — this article will explain how vinyl films are made, outline their characteristics and offer suggestions to help you determine the right film for the situation.

Roll the film

Pressure-sensitive films typically comprise facestock, adhesive and liner paper. Polyvinyl chloride (PVC) films comprise the majority of sign materials. Polyesters, foils and reflective films are also used for sign applications, but to a much lesser extent.

PVC films include both cast and calendered materials. Because manufacturers employ different processes and raw materials to produce these films, they exhibit unique physical properties and performance characteristics.

Calendered films (which some in the sign trade still call "4-mil" vinyl) are available in various gauges. However, today, it's doubtful that any are actually 4 mils thick. As technology improved, films got thinner.

Raw materials for calendered films — PVC resins, pigments and plasticizers — are usually solid, but may also include some liquid ingredients. After being measured and fed into a mixer, the materials are blended to ensure the finished product's color consistency and uniformity. These materials are then heated and rolled into a continuous film sheet.

The finish on the roller's surface determines the finish on the film's surface. A highly polished roller finish,

These cool graphics were fabricated using Avery Graphics' (Painesville, OH) MPI 1005 high-gloss, cast vinyl and protected with a Nazdar (Shawnee, KS) clearcoat. Cast-vinyl manufacturers typically use higher-grade pigments and more stable plasticizing oils, which make this material more comfortable than calendered film.

for example, imparts a high gloss. The calendering process is like rolling out pie dough with a rolling pin. This stretching process builds mechanical stress into the film. Just as dough shrinks after rolling, calendered vinyl contracts immediately.

For some applications, calendered vinyls are a suitable, cost-effective alternative to cast vinyls. Cheaper raw materials, such as lower-grade pigments and plasticizers, mean lower cost. However, they may compromise performance.

Plasticizing oils that comprise calendered film are much more volatile than those used in cast vinyl. When subjected to UV light, heat, chemical spillage and pollution, less-stable plasticizers can leech out of a calendered vinyl and migrate into the film's surface or adhesive.

Loss of oil causes vinyl to lose flexibility. When this happens, brittleness and cracking usually occur. Plasticizer migration into the film's adhesive can soften it and destroy its adhesive bond, resulting in either adhesive failure or graphic shrinkage.

When dealing with large quantities, the calendering process outperforms casting. A high-speed, high-output manufacturing process, calendering produces large volumes at a relatively low cost. A typical production run can produce up to 225,000 sq. ft. Because mass production increases inventory investment, vinyl manufacturers limit the color range of calendered vinyls. With such high production requirements, custom colors are often unfeasible.

Calendered films possess other advantages. Excellent tensile strength enables calendered film to pull itself through the extrusion process. Thus, it's less likely to tear than cast film, and it can withstand greater impact and abrasion. This is why thicker calendered films are sometimes used to cover gas-pump skirts. Their stiffness and strength also facilitate application and film removal.

Calendered films' dimensional stability, flexibility, caliper control and durability have improved, but they still shrink. Regardless of film thickness or the degree of applied heat and effort, calendered films tent around rivets and lift in corrugation valleys over time.

Honey, I shrunk the vinyl!

I formerly thought all calendered vinyls were created equal. Boy, was I wrong! While working for a screenprinter, I'd been using a high-end calendered film that was nearly as costly as cast films to decorate building and canopy fascias for gas stations and convenience stores. The manufacturer also marketed another calendered vinyl that cost approximately half the price of our standard material. I thought the two products couldn't be that different.

I decided to experiment with the cheaper film for a multi-color, truck-door decal. We printed one color, letting the ink air-dry overnight. The next day, we discovered the vinyl shrunk 1/4 in. away from the liner all the way around the decal. As a result, the second color was impossible to register.

We scrapped my little experiment and started from scratch. This taught me the difference between monomeric and polymeric films.

Polymeric films comprise more complex building blocks with a higher molecular weight than simpler, cheaper monomeric films. The complex structure stabilizes polymeric films and contributes to increased durability.

Liquid origins

Cast films — typically called "high-performance" or "2-mil" vinyls — comprise liquid materials, rather than solids. The liquid, which resembles paint, is called an organosol. Similar to calendered films, cast materials contain pigments, plasticizers, PVC resins, solvents and additives.

However, cast-vinyl manufacturers use more costly, automotive-grade pigments and heavy, stable plasticizing oil. The film retains the weightier oil throughout its life, which keeps the film flexible. By comparison, calendered film's lighter, more volatile oil helps lubricate extrusion rollers during the manufacturing process. However, eventually, the lighter oil leeches out and leaves the film brittle.

When manufacturing cast films, liquid organosol is coated or poured onto the

This Toyota Celica GT was decorated with Oracal USA's (Jacksonville, FL) Series 851, premium-cast vinyl. Such films apply more readily to corrugations, rivets, and curved or uneven surfaces than calendered vinyl.

casting paper. As the paper travels through a 75-ft. oven, the organosol is baked at 400°-450°F.

Baking evaporates solvents from the wet organosol, leaving a dry film. In this "fusion" process, the plasticizer bonds with the PVC resin, giving cast-vinyl film its strength.

This fusion is the key to making film, because it affects virtually all its properties, such as elongation, modulus (measurement of the stress a material can absorb before losing its elasticity), adhesive anchorage to the facestock, ink adhesion, system durability, shrinkage, and the film's hue and gloss.

The bond's quality depends on the molecular weight of the plasticizers, PVC resin, pigment and additives, the type of solvent used, and the length of the curing process.

Little mechanical stress is created during the casting process because the organosol merely lies on the surface of the casting paper while the film is fabricated. Consequently, film shrinkage is low.

Cast vinyl's low tensile strength contributes to the film's stretchability, and consequently, its conformability. Therefore, cast film can cover such demanding substrates as rivets, corrugations and textures.

Cast vinyl is produced at much slower speeds and in much smaller batches than calendered film. While this adds to its cost, cast vinyl can be produced in various colors, including batch lots as large as approximately 10,000 sq. ft.

These boots were made for walking, and this floor graphic will have to withstand the wear and tear. The graphic comprises Drytac Corp.'s (Richmond, VA) VersaJet adhesive-backed, calendered film; the company's MediaShield Emerytex overlaminate protects this big burger. While calendered films have greatly improved in recent years, they're still known as intermediate vinyls, whereas cast materials are called high-performance.

Shop and compare

Film manufacturers often force signmakers to determine a vinyl's suitability for a particular application, and then hold them responsible for problems. However, you might not have a lab outfitted with up-to-date testing equipment.

Several criteria arise when making your material selection. Always study the manufacturer's product data sheets and other technical information. Generally, cast vinyl is referred to as high-performance, and calendered vinyl is known as intermediate material.

But, buyers must beware and take what some manufacturers publish about their products with a grain of salt. Some manufacturers play name games. For example, a "high-performance" label doesn't necessarily mean the film is cast vinyl.

When it's too warm to cook, and you crave a ham and swiss on pumpernickel, this sign — made from Arlon's Calon Series 2100, 2-mil, cast material — lets you know where to go. When working in a warm, humid environment, note how flat the film's liner will stay, because it expands or contracts as it gains or loses moisture.

When conducting your tests, research the following:
Vinyl manufacturers perform cutting and weeding tests
to evaluate their own and competitors' products. Cut letters of various sizes, such as ¼, 1 or 2 in. Some manufacturers cut the entire alphabet, whereas others cut letters only in the top row of keyboard keys.

Conduct your cutting and weeding tests side by side for easier vinyl comparison. You'll find that the facestock's brittleness or softness may impact cutting. Further, colors in a vinyl series may cut differently than others. Facestock's caliper variations can also complicate cutting. Often, cast vinyl cuts more easily than calendered, because it's thinner and provides better caliper control during manufacturing.

When testing a vinyl, pay attention to its weedability. To a large extent, a film's release characteristics determine how well it weeds. Release values, which many vinyl manufacturers publish, measure the force required to remove a release liner from a pressure-sensitive material, including vinyls.

Without standardized industry procedures, test results from one lab to another would vary widely and have little meaning. Vinyl manufacturers conduct their tests according to accepted industry standards, such as those created by the Pressure Sensitive Tape Council (PSTC) and the Tag & Label Manufacturers Institute (TLMI).

Remember that release values can gradually grow. This happens because time and environmental conditions can wear on the liner and facestock, requiring more force to separate them.

To gauge this, testers commonly use a TLMI test instrument. Using this instrument, the facestock of a 2 × 10-in. material sample is attached to the sled of a release tester, while the sample's liner attaches to the tester's stationary arm. As the sled travels its track at 25 ft. per min., the test equipment measures (in grams) the force required to peel and separate the sample's two parts.

This trailer's graphics were fabricated using Calon Series 6000 2-mil, cast material. When testing the weatherability of such material, Mother Nature will tell the tale as effectively or better than a UV or xenon testing lab.

The tester may calculate a release value of 100gm per 2-in. width after coating. One week later, the reading could measure 140gm. In a few months, the value may grow to 160gm.

When evaluating vinyl, make sure the liner paper lies reasonably flat; excessive edge curling makes handling more difficult. High release values — such as those in the 200- to 500-gram range —
indicate poor graphic transfer from the liner, making application difficult. Low release values, such as 60-70 grams, can indicate poor film stability on the liner, which causes problems for screenprinting, die-cutting and plotter-cutting applications.

If you're working in a humid environment, such as Florida or Houston, you'll want to note how flat the liner stays. As it gains or loses moisture, liner paper can expand or contract. Vinyl manufacturers measure these changes and the paper's moisture content. Liner growth can cause curling, which makes plotter cutting difficult.

Test, don't guess

To assess a vinyl's durability, vinyl manufacturers will perform a variety of tests, including outdoor weathering at perhaps a Florida or Arizona test site, as well as simulated weathering tests in a weatherometer, such as a UV tester — for instance, an accelerated-weathering tester from Q-Panel Lab Products (Cleveland) — or a xenon test chamber.

A UV tester exposes vinyl to fluorescent UV rays. Although UV light represents only 5% of the sun's rays, it causes the most damage. This type of testing measures worst-case environmental scenarios, and spectrophotometers gauge color weathering after 500, 1,000 and 2,000 hours of exposure under such intense conditions.

Xenon test chambers expose samples to both infrared and UV light using a filtered, xenon arclight. Although more accurate than UV testing, this method is considerably more expensive.

The results obtained with test equipment often don't correlate with real-time outdoor exposure.

A sample can fail in a QUV tester and pass in the real world, and vice versa.

Very few sign companies and screen-printers can afford expensive weathering equipment, but that's OK. Nothing beats the time-tested method of exposing a sample to the elements. Mother Nature can tell you more about how a vinyl performs than costly equipment. A friend in Minnesota conducts his own weathering tests by applying vinyl samples to banners he hangs outside his shop.

Still, test equipment is useful as a pass/fail exam to screen new vinyl formulations. Weathering tests measure gloss and color loss visually, as well as tangibly, with a gloss meter. These tests compare weathered samples against a control sample.

Technicians also look for water spots, chalking, dirt retention and corrosion between the adhesive and substrate (and the film, if it's either a reflective or metallic specialty vinyl). In evaluating a vinyl film's dimensional stability, the technician will cut an "X" in the center of the sample and measure the shrinkage from the cut line.

Calon Series 2500 2-mil, translucent, cast vinyl enables this shopping center to tout its tenants. Sunlight's potency and impact on such an outdoor installation will vary based on climate and altitude.

Manufacturers often test new material against "aged" vinyl that's been baked in an oven for a specified period, such as seven days at 120°F.

Heat aging simulates the effects of time and demonstrates how storage temperatures affect the film's physical properties. For example, aging tests may reveal a severe reduction in elongation (how far a material will stretch before breaking) and tensile strength (the amount of applied force required to break a film).

Eleven queries

To determine the best vinyl for your application, ask the following questions:

- For what type of sign will the vinyl be used? Banners, vehicle graphics, interior graphics, exhibits or displays have distinct vinyl needs.
- To what substrate will the vinyl be applied? Adhesion values differ greatly based on the substrate's paint system. If the vinyl is applied to rivets, corrugations and textures, it should be conformable and not cause tenting or lifting.
- What is the substrate's condition? When you conduct a survey, check for rusting, body damage and other signs of deterioration.
- Will the graphics be installed indoors or outside? The sun's bleaching rays vary in strength in different parts of the country and at different altitudes. At higher altitudes,

thinner air offers less protection from sun damage. In any outdoor location, pollution can damage the surface of vinyl graphics or harm adhesive at the film's edge.

- How easily does the film cut and weed? How does vinyl perform when combined with other graphic components used in a vehicle-marking system, such as inks, clearcoats and overlaminates? Clearcoated or laminated graphics withstand Mother Nature's wrath much better than those left unprotected.
- Will your vinyl perform well enough? Know the job's durability requirements.
- What are the economic factors, and who are your competitors? Your customer wants to buy with confidence and be assured of a high-quality product that looks good at a reasonable price.
- Will the graphics be exposed to solvents, chemicals or abrasions?
- What type of cleaning chemicals will be used, and how frequently will the graphics be cleaned? Some cleaning procedures and products can actually harm graphics.
- How easily does the film handle during application?
- How easily can the graphics be removed? Will it be just your average, mind-numbing, pain-in-the-neck job? Or, will it be a disaster of biblical proportions that will make you rue your decision to become a signmaker?

Clearly, the above list doesn't address every circumstance. To cover all the bases, ask questions, visit job sites and anticipate problems. In many cases, calendered products will meet your needs for applications that span less than five years. For long-term signage (five to eight years), cast vinyl may offer significant advantages.

Chapter 12
Calendered Vinyl

Because technology continually evolves, don't allow your first impressions to be lasting ones. Unfortunately, I haven't always heeded my own advice. Regarding calendered vinyls, some prejudices have prevented me from embracing this film technology's advances.

A recent tour of American Renolit Corp. (La Porte, IN), a modern plant that manufactures calendered vinyl, has altered some of my existing beliefs. Thus, I'll discuss the calendered-film manufacturing process and amend some of my previous disparaging comments. Also, I'd like to thank John Gooch, American Renolit's vice president of sales and marketing, for helping me prepare this chapter.

Chicago-based Thomas Melvin Painting Studio used ORACAL® 8300 Transparent Cal to complete a window mural that scaled 32 x 108 ft. The transparent, calendered film features a 3-mil thickness, five-year durability and a clear, permanent adhesive.

It was 20 years ago today...

Twenty years ago, calendered films were classified as either monomeric or polymeric. Today, a broader selection of resin grades and plasticizers has expanded that range. The new list includes: monomeric calendered (3.0 mil thick), blends (3.0 to 3.2 mil thick), standard polymeric calendered (3.0 to 3.2 mil thick) and high-performance polymeric calendered (2.4 mil thick).

Monomeric calendered vinyl is a low-cost alternative (keep in mind, you get what you pay for). Because the film uses a lower, molecular-weight monomeric plasticizer, this formulation is more suitable for indoor and short-term, outdoor applications.

Standard polymeric calendered vinyl utilizes a higher molecular-weight polymeric plasticizer system, which makes the film suitable for outdoor applications.

Next, there's high-performance calendered vinyl. The combination of a thinner gauge (2.4 mil), and a polymeric plasticizer system, provides an excellent product that's conformable enough for applying over demanding surfaces — such as rivets — in outdoor applications.

I'm sure some industry purists think "high-performance calendered vinyl" is an oxymoron. However, cast vinyl isn't the only high-performance film available.

Indeed, high-performance calendered material isn't the same as cast vinyl. However, its performance is similar. Plus, measuring at 2.4 mil thin, the new generation of calendered films is nearly as thin as cast material. Many of these films also have an exceptional gloss level.

Although the differences among the various categories can be confusing, high-performance, calendered-film usage as a cost-effective alternative to cast vinyl films has grown. For a wide range of signage applications, high-performance calendered films can satisfy a signmaker's needs.

Finally, "blends" — cost-effective vinyl that provides good outdoor durability for general applications — result from the combination of the two previously mentioned, main plasticizer systems.

An associate who works for a large coating company recently subjected samples of monomeric vinyl, a standard polymeric film and a blend to heat-aging tests. As expected, the monomeric sample had less dimensional stability. Remarkably, however, the blended product exhibited nearly the same dimensional stability as the standard polymeric vinyl.

Blends are a quality, low-cost face-stock and an excellent choice for less demanding, outdoor applications.

Full polymeric formulations are recommended for more demanding applications in high-solvent environments, such as gas-tank decals and Arizona inkjet printing.

Calendered-vinyl technology improvements have helped increase productivity, improve product quality and lower costs for signmakers and screenprinters. Some improvements include thinner films, improved gloss levels and better gauge control. Variations in sheet thickness are less than 5%.

Years ago, production was limited to long runs of a single color. Hence, fewer colors were produced. However, today's film manufacturers can change colors on the fly.

Raw materials

I once compared the calendering process to rolling out pie dough. However, the manufacturing process is more complicated. In the calendering process, PVC material is squeezed between gigantic, heated, polished-steel rollers that form the vinyl into a sheet.

A modern calendering line is also more expensive than grandma's rolling pin, with capital equipment investments ranging typically from $10 million to $15 million. In addition, the calendering rolls alone cost several hundreds of thousands of dollars.

The production line I saw featured advanced computer control, which allowed continuous monitoring and in-process adjustments of machine functions.

In addition to focusing on the raw materials that create calendered vinyl, I'll review the calendering process's basic steps: blending, fluxing, milling, calendering, embossing, cooling and winding.

PVC resin is the key ingredient. However, by itself, it's a very hard and brittle plastic material. Additives — plasticizers, stabilizers, lubricants, pigments and processing aids — change the film's physical properties and make it easier to process.

Plasticizers, which are liquids incorporated into the formulation, soften the hard PVC resin and make the films more flexible.

Plasticizers in polymeric calendered vinyls are more complex than those in monomeric films. Monomeric plasticizer comprises linear (sometimes branched) molecular chains, whereas polymeric plasticizer features complex branched chains of a higher molecular weight. The weightier polymeric molecules resist migration and create a more stable and longer-life plastic.

Plasticizers add flexibility to the final film. Plasticizer additions soften the film's feel, or "hand." Normal vinyl films used in graphics applications contain between 20% and 25% plasticizer. The higher-molecular-weight, polymeric-plasticized mixes have a higher viscosity (rheology) than monomeric systems, and are processed at slower production speeds. These plasticizers are more expensive than the monomerics, and are less efficient, which means they require more content of a more expensive material manufactured at lower speeds.

The plasticizer's molecular weight helps determine its stability. Higher-molecular-weight, polymeric plasticizer comprises very big, bulky, slow-moving molecules. Hence, they stay in the film.

The early, lower-molecular-weight monomeric plasticizer molecules were more mobile. They moved around easily and readily migrated out of the PVC into other adjoining materials, such as the adhesive coated on the film.

Modern monomeric plasticizers are less mobile, which limits migration into the adhesive.

Heat and UV light prematurely age products, especially vinyl. As the film ages, it yellows and degrades. Films can lose their elasticity and become brittle.

American Renolit Corp. is a La Porte, IN-based modern plant that manufactures calendered vinyl. John Gooch, the company's vice president of sales and marketing, helped prepare this chapter.

Heat and UV stabilizers slow down this aging process. Although monomeric formulations don't usually contain UV absorbers, they're often incorporated into the vinyl (and adhesive) — for overlaminating applications — to impart some UV protection to the digital image or photograph being protected.

Lubricants assist the film's release from the hot calender rolls during production and act as internal processing aides during production. They not only prevent PVC film from sticking to the rollers, they also improve the compound's process.

Pigmentation is achieved via inorganic and organic pigments, which are usually first ground into the plasticizer to achieve a specified particle size (for full-color development). The pigment selection will depend upon the vinyl's end application.

Economical inorganic pigments provide higher opacity than organics. High-performance calendered vinyls use the same pigment opportunities as the more expensive cast products.

The calendering process
By itself, the PVC resin is as hard and brittle as a saltine cracker. Thus, to transform the PVC resin into a flexible plastic, it must be compounded with other vinyl-film ingredients.

Once all the raw materials are weighed and blended, into a very fine powder with the consistency of cake flour, the blend is fed into the extruder's screw — the machine's fluxing section. Under heat and pressure, the extruder continues the mixing process, evenly dispersing all the additives with the PVC resin. In the fluxing or plastification process, all the separate ingredients fuse together into one homogenous mass of plastic called the "melt." As the fine powder melts, at approximately 300°F, the extruder kneads the material into a hot, twisted plastic rope. The extruder also helps strain out any foreign particles, which could damage the machinery.

Next, the hot, plastic rope is fed into a two-roll mill bin and rolled into a rough sheet of film. In the manufacturing process, the heat is continually increased, making the film more malleable so it can be rolled increasingly thinner. During this milling process, edge trim can be reworked into the mix.

After the two-roll mill, the material passes a metal detector. This inspection step prevents any metal from reaching the calendering rolls. Even the tiniest metallic speck could cause irreparable damage and necessitate roller replacement.

Calender rolls
The calender comprises four, rather heavy, highly polished, 2-ft.-diameter steel rolls. Calendering rolls can be arranged into various configurations, such as "L," "F" (or inverted "L") and "Z". The plant I visited utilized a four-roll "L" configuration.

During the calendering process, the heat increases, and the vinyl sheet passes between the rollers. During this process, the film is squeezed into a much thinner and wider sheet.

The calender rolls subject the vinyl sheet to thousands of pounds of pressure per square inch. Such intense forces and rapid production speeds can bend and deflect these massive rolls. To compensate for the rolls' deflection, some complex mechanical-engineering ideas have been incorporated into modern machinery, including crowned rolls (which are thicker in the middle than on the edges) and rolls that are crossed slightly to one another, rather than perfectly parallel, and designed to counter the rolls' bending by applying pressure in the counter-direction.

These measures ensure an optimum profile so the vinyl's caliper remains uniform. After the web travels through the production line's calender section, the strip-off, or pick-off roll, strips the sheet from the calendering rolls.

To impart the film's surface finish, the sheet goes through an embossing unit. Here, the film is pressed onto an embossing cylinder with a matte-rubber pressure cylinder. The resultant surface finish depends upon this embossing cylinder's condition. A high-gloss surface is achieved from a highly polished chrome cylinder, whereas a matte surface is achieved from a matte-engraved emboss cylinder. The film's reverse side has an unspecified matte surface.

As the vinyl sheet passes over and under a series of chilling rolls in the cooling section, the vinyl quickly cools. The cooling process anneals the vinyl into its final form.

The vinyl film is then wound, using highly sophisticated, progressive-tension winders, to minimize any tension that could result in future, dimensional-stability issues. For high-gloss films with a soft hand, it's important to keep roll lengths to a minimum, with controlled winding tension, to reduce the tendency for gloss reduction through the roll. This reduction is caused by "cold embossing" — the matte reverse side presses onto the roll's high-gloss surface. This is temporary "damage," and the gloss can easily be refreshed with heat during subsequent processing.

Although the facestock is a vinyl film's key ingredient, it's only one ingredient in the pressure-sensitive sandwich. How a film cuts, weeds and handles depends on how all the components work together. Two rolls of the same facestock, coated with different adhesives, will likely perform differently. Remember this when you evaluate a new film product. And, of course, keep an open mind.

Chapter 13
Paint-Mask Vinyl

Why paint graphics when using vinyl is easier? In truth, paint is more economical than vinyl in some cases. This is especially true when the graphics cover a very large area. I know this sounds like a sacrilege coming from a vinyl-graphics columnist, but that's the way it is.

Paint and vinyl are generally competing decoration processes, but they can be used together, such as on this tour bus decorated by Performance Graphics (Elkhart, IN). When paint is preferable to vinyl graphics, use quality vinyl mask to mark off areas to be painted.

Here's a time-tested method to faithfully reproduce a design using paint. Some old-school signpainters use opaque projectors to transmit their design onto a paper or plastic sheet. After drawing the design onto the sheet, the painter uses a pounce wheel, a tool with very sharp points that punches small holes into the sheet. Then, he tapes the sheet into place.

Tapping a pad filled with graphite — a pounce pad — transfers the pattern. Using the pounced pattern, the fabricator handpaints the graphics. This archaic technique is still useful.

Paint-mask vinyl

For painted graphics, you can also use paint-mask vinyl. Paint mask is available in various colors — white and yellow are the most popular colors in the United States. Yellow mask generally performs better; it can withstand high curing temperatures during baking cycles. In other countries, vinyl-graphics fabricators use transparent paint mask.

After cutting the film in a plotter, the user reverse-weeds it to create a stencil. Using vinyl with a low-tack application tape, the stencil transfers to the substrate.

Some signmakers will use cheap, calendered vinyl instead of paint mask. The materials aren't the same. If the film and adhesive don't remove cleanly, you'll waste time removing adhesive and fixing mistakes.

When using a laminator to adhere the application tape to the paint mask, apply low pressure (10 to 15 lbs.). Heat from thermal laminators, which varies from 90° to 100° F, helps the application-tape adhesive to flow out, especially in the winter, when shops can be cold.

When applying very large sections of cut paint mask, use a heavyweight, premium application tape with a very low-tack adhesive. A few application tapes are designed specifically for applying paint-mask vinyl. If you don't know which tape to use, ask your distributor for a recommendation.

The application tape needs enough tack to transfer graphics, but shouldn't pull the paint mask from the substrate during tape removal. Many painters prefer heavier application tape because it's less likely to tear when laminating the tape to large sheets of graphics. If the tape tears, and falls onto the graphics, the vinyl stencil is trash.

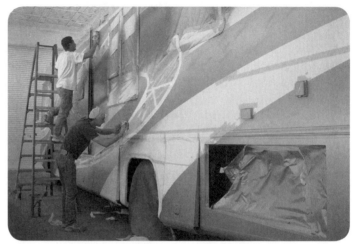

Finally, either spraypaint or roller-paint the graphics. Using paint mask is faster than pouncing patterns. You can also brush on the paint. If you add a flow enhancer, such as Penetrol, you will eliminate unsightly brush marks.

Due diligence

Selecting the right paint-mask vinyl for your shop is critical. Many paint masks are on the market, and they're not all created

Performance Graphics installers squeegee paint mask onto the bus surface. Some installers use low-end, calendered vinyl, which can leave adhesive residue that requires cleanup and slows production.

equal. Before purchasing, test and evaluate the paint mask before production.

Some common problems that signmakers encounter with the wrong paint mask include adhesive delamination, edge lifting and "fish eyes," which are tiny circular craters in the paint or clearcoat that form over a substrate contaminant.

When qualifying a paint mask, apply some to a test panel, spraypaint the mask as you normally would and watch what happens. If the painted graphics will be oven-cured, simulate the heat they'll encounter.

Curing times and temperatures depend on the paint type. For example, one paint system we recently tested cured at 150° F after 20 minutes. However, some paint systems require higher temperatures and longer curing times, whereas others need only air drying. Test the paint mask with every system you use so that all components are compatible.

Performance Graphics outputs the pattern onto its plotters, and then reverse-weeds the film to create a stencil. Next, the installers transfer the stencil to the substrate.

If you apply paint mask to a painted surface, the paint must be thoroughly cured. If it doesn't, the paint's solvents will likely outgas and react with the mask's adhesive. These solvents can turn the adhesive into a gooey mass that leaves a residue when removing the paint mask.

Installing paint-mask film is much easier than applying regular vinyl. If a wrinkle or bubble appears in the film, don't worry. Mistakes repair easily with a little masking tape.

When painting vehicle graphics, just apply the paint mask over obstructions. After carefully removing the application tape, squeegee all paint-mask edges again.

Application tape

Application tape is self-wound.

The tape's paper backing picks up minute amounts of adhesive. As you squeegee paint mask in place, you may potentially transfer some adhesive to unmasked areas, which, if left, contaminates the paint. To prevent this, use a low-friction sleeve on your squeegee. Like a magnet, the sleeve picks up adhesive particles and prevents substrate contamination.

Be sure the mask's edges fit tightly against the surface. If it doesn't, paint will blow back underneath. Where the vinyl lifts or tents at the edges, slit the film with an X-acto® knife. Be careful not to cut the substrate surface.

After reapplying the paint mask, use only premium masking tape to clean up any dirty edges. It makes no sense to save a few cents buying a cheap tape; there are no shortcuts to quality and value. Some cheaper masking tapes allow paint to bleed through.

After applying the paint-mask vinyl, sand the substrate surface with 500- to 1,000-grit sandpaper before painting. Sanding roughs the surface, promotes good paint adhesion and minimizes defects. Use an air hose to blow off paint dust. Then, wipe the surface clean using a tack cloth.

Also, the paint-mask vinyl may pick up silicone from the release liner and transfer it to the painted substrate. If you clearcoat the paint after masking, fish eyes can appear wherever silicone contaminates the paint.

Although silicone particles commonly cause this problem, particles of dust, wax and oil may also contaminate the surface. That's why the substrate must be perfectly clean before paint or clearcoat applications, especially if what's painted is very expensive. Repainting to fix a mistake is a costly remedy.

Chapter 14
Translucent Vinyl

Photo courtesy of Avery Dennison

Most people don't like the spotlight, where warts are prominently displayed. The same philosophy applies to translucent vinyl. When the lights are turned on, inconsistencies in color and film gauge appear. In this chapter, I'll cover translucent-film selection, design and application options.

Cast or calendered

Translucent-vinyl applications aren't limited to backlit, building signage. Other uses include taxi-top signage, mall and airport-terminal advertising, menuboards and canopy fascias.

Consider these points when deciding whether to use cast or calendered translucent films:

Inkjet-printed, translucent vinyl, such as this multi-display, health-club program, typically needs an over-laminate to counteract contaminants and UV damage.

- What's the application? For less demanding applications, such as menuboards and directional signs, a calendered translucent vinyl suffices. For long-term, outdoor applications, I would choose cast films.
- What are the durability requirements?
- What are your customer's expectations?
- What is the budget for the job?

User preference dictates the choice between lower-cost, calendered vinyl versus a premium, cast film. For most applications, you can't go wrong with cast vinyl.

However, that doesn't preclude calendered-film usage. Newer generation, calendered, translucent films feature such improvements as color consistency and gauge control. Also, some calendered films offer five-year durability.

Because calendered films cost less than cast, they've become the film of choice for a wide range of backlit signage. If you use a calendered vinyl, buy one designed for the job. Although calendered films are translucent, not all of them are suitable for backlit signage. Lesser calendered films may look fine during the day, but inconsistencies in color and film gauge readily appear when the sign box lights up.

I'm not saying calendered films are a bad choice for backlit applications. Better varieties have standardized thicknesses. Light-transmission discrepancies cause film-gauge variations, which appear when illuminated.

If color consistency deviates, film can exhibit blotches when illuminated. Plasticizers in cheaper calendered films can migrate to the film surface. If you screenprint or digitally print the film, plasticizing oils inhibit good ink adhesion.

In most cases, an overlaminate should protect the print. Inks tend to be fragile, and overlaminates neutralize such hazards as UV light, air pollution and bird droppings.

"Translucent films must be able to withstand high temperatures," Chuck Bules, Arlon's (Santa Ana, CA) technical services manager, said. "On a hot summer day, temperatures inside a light box can easily exceed 150° F. When subjected to baking heat, cheaper calendered films can shrink and crack."

Thermoforming

Thermoforming poses a challenge for translucent films because exposing vinyl to high heat for a prolonged time can damage the film.

"In the thermoforming process, not only is the vinyl subjected to high heat, but it's also stretched in every direction," Bules said. "The film must have extensive elongation and still retain its color. That's some trick. Color distortion, gels, streaks and pinholes are unacceptable in backlit applications."

In addition to high-temperature durability, translucent films require a weedable adhesive when large film sections are applied to plastic sheet, then cut and weeded. Cutting vinyl that's been applied to a plastic sign face can damage the substrate. Avoid overcuts, which will widen and show up under light. To aid weeding, warm the vinyl slightly with a heat gun. This will soften the adhesive and allow easier removal.

Translucent vinyl can be applied either to the plastic sheet's first or second surface. During decoration, the vinyl film shouldn't touch the mold to prevent vinyl damage.

Because vinyl may be applied to heavily plasticized, flexible-face and awning material, the adhesive must also withstand the plasticizing oil. "Worm tracking," a common vinyl failure on awnings, occurs after the plasticizer softens the adhesive. As the vinyl expands and contracts on the substrate, tiny tunnels form underneath the film.

Matching

Because backlit graphics encounter myriad environmental stresses, all components must be compatible. Components include the vinyl, substrate, ink system and overlaminate. Your distributor can help you select the right components. Also, make sure wet applications don't discolor the adhesive.

For short-term, promotional graphics, consider how easily the film can be removed. Calendered films are usually easier to remove than cast films. Because of its high tensile strength, a calendered film peels from the substrate in large sections, if not in one piece.

Design considerations

Prior to the design stage, conduct an onsite survey. Pay careful attention to any building obstructions that would cause installation problems. Also, take plenty of pictures.

In my opinion, signs with dark backgrounds are much more readable than signs with white backgrounds. Avoid using thin letters against a white background or outline. When a sign is illuminated, the light tends to bleed around the letters' edges and make copy more difficult to read. Colors should be viewed in all conditions.

Non-vertical awning surfaces can't display any upper graphics, and suffer twice the UV-light intensity. Vinyl applied to horizontal surfaces fade first. In northern climates, snow accumulation can create vinyl edge lifting.

Translucent vinyls are engineered for acrylic, polycarbonate and flexible-face materials and glass. In most cases, either the sign cabinet or awning is backlit. For this reason, color consistency is critical, especially when one film sheet overlaps another.

Distance between fluorescent lighting and the sign face is also critical. If the lighting is too close to the graphics, hot spots can appear. I once made the mistake of substituting a flat face for a pan-formed face. What a gaffe — the lamps behind the face were too visible.

If the face is extended a couple of inches, pan-formed, plastic signs reduce the likelihood of hot spots. Also, pan-formed faces are stronger than flat sheets.

Using a diffuser film on the face's second surface also disperses light and prevents hot spots from washing out the image. Diffuser films are rated on the amount of light transmitted thought the film. Today's films provide 30% to 70% light transmission. Remember, the diffuser film can darken the image slightly.

Light-enhancement films, when applied inside a sign cabinet, reflect light off the surface and distribute uniform light, which also minimizes hot-spot occurrences.

Vinyl application

Even if a plastic sheet includes protective masking, clean the sign face before applying the graphics. Before cleaning any substrate with a solvent or detergent, carefully read the manufacturer's instructions. For most plastic-sheet applications, the most reliable cleaning method is detergent and water. Solvents subject plastic sheets to chemical stress, and can cause cracking.

If you're decorating a polycarbonate sheet, remember that this plastic absorbs air humidity — it's "hygroscopic." As the plastic outgases moisture, it can cause bubbling on any applied-vinyl graphics.

As a rule of thumb, polycarbonate sheets should be dry before decoration. To check for outgasing and ensure vinyl-ready material, apply a piece of vinyl to a polycarbonate sample, and bake it in an oven at 150° F for a day. If bubbling occurs underneath the film, re-dry the sheet.

Some professional decal installers only apply vinyl dry. In most cases, I agree that dry application is preferable. However, applying a translucent vinyl to a plastic sign face is the exception to the rule, because the sign face tends to attract vinyl like a magnet. For this reason, I believe wet applications serve such jobs better.

Photo courtesy of Arlon Inc.

Light transmission is an important consideration for translucent films. Color or thickness discrepancies will make illuminated film appear blotchy.

You can make your own application fluid by mixing 1/2 teaspoon of a dishwashing detergent, 1/2 teaspoon of isopropyl alcohol and 20 oz. of water. I prefer using a commercial application fluid, such as Rapid Tac, Window Juice or Splash.

Use the minimum amount to accomplish the task. Lightly mist the surface with the fluid; an excess can create unwanted residue. If the liner gets wet, the silicon layer can flake off and contaminate the adhesive. When the sign is illuminated, these areas appear as dark blotches.

If you make your own concoction, don't use dishwashing liquid that contains moisturizing lotions, which contaminate the vinyl's adhesive and create failure.

When applying graphics, start in the center. Apply enough squeegee pressure to force any application fluid from under the vinyl, and always use overlapping strokes. After removing the application tape, re-squeegee the entire graphic using a squeegee covered with a low-friction sleeve. This protects the vinyl from scratches.

When working with translucent films, avoid seams if you can. When translucent films overlap, the seam will be noticeable when the sign box is illuminated. Of course, overlaps aren't always avoidable, but keep them to a minimum. Seams generally shouldn't exceed 1/16 in. Abutting films can shrink and leave a noticeable gap when the sign is lit.

The film's light and color transmission can vary from roll to roll — sometimes, even within a single roll. Therefore, overlap films from the same roll or lot number. Even then, color variations and inconsistencies can occur within the same roll. For this reason, if a film needs to be seamed, take the time to match colors to ensure consistent appearance.

Chapter 15
Reflective Vehicle Graphics

When it comes to fleet graphics, my heart pumps faster when my headlights illuminate a trailer with well-designed reflective markings. I admit, maybe I should get a life. However, I've seen similar reactions from middle-aged businessmen who spend 30 minutes turning their high beams on and off to experience new reflective truck graphics.

That's the level of interest and excitement that great graphics should create. A winning, reflective-graphics program effectively

Reflective vinyls — such as 3M's™ Scotchlite™ 280i pictured here — need to have a high candle-power rating for ideal efficiency.

communicates the client's fleet advertising message while providing optimum safety at a competitive price. That's much easier said than done.

Over the years, I've worked with many talented pros who've taught me the essential do's and don'ts in the design and manufacturing of reflective fleet markings. In reviewing some of the following time-tested tips, I hope you discover new ways to improve your designs and give your customer the biggest bang for his reflective buck.

Reflecting on reflection

Approximately 18 years ago, I was trying to win over a large tanker fleet account on the south side of Chicago. The graphics package incorporated green and blue reflective markings on the sides and rears of the units.

Fortunately, my employer was practically the best production man in the business. After carefully inspecting the printed markings, Larry noticed that one of the colors didn't reflect. He was sure that white had been added to the ink system to achieve a color match. Mixing any opaque color with transparent ink will kill reflectivity.

After I demonstrated to the fleet manager that he'd been buying reflective markings that didn't reflect, he was furious. In his mind, he'd been cheated. That signaled the end for the incumbent printer.

Whether you screenprint or digitally print colors, make sure that the ink neither clouds the nighttime appearance of color nor deadens the reflectivity. Surprisingly, colors printed with some thermal-resin printers — such as the Gerber Edge™ — can have higher candle-power ratings than pigmented sheeting's equivalent. However, be sure to print with transparent foils.

The moral of this story? If you quote on an existing reflective-graphics program, take the time to inspect the markings at night. A spotlight is a handy tool for nighttime viewings. As you evaluate the graphics program, ask the following questions:

- What are the existing program's shortcomings or problems?
- How can you re-engineer the job to improve the nighttime visual impact of the graphics?
- How can you improve the graphics' readability?
- How could you make the program more cost-effective?

Remember, the key to achieving better graphics solutions is to ask better questions.

Do your sheets match?

Early in my career, a co-worker learned a costly lesson about the importance of matching reflective sheets for daytime and nighttime appearance. The design for an Ohio-based manufacturer incorporated a 7-ft.-tall graphic of a staple gun. To minimize material waste, he printed the pictorial on two sheets of reflective material, with one sheet coming from a 3-ft.-wide roll and the other from a 4-ft. log.

However, the two white reflective sheets were noticeably different. During the day, one sheet appeared yellowish, while the other looked light gray. The nighttime variance was even worse.

Using material from two different rolls was the first in a series of errors. Color and reflective values often vary from one roll to the next. That's why it's prudent to use material from the same roll. If that's not possible, at least make sure all the material comes from the same lot number.

Second mistake: Each log used for the staple-gun pictorial came from a different production run. Reflective sheeting's color can vary greatly from one lot number to another, and also from one roll to another within a specific lot. Appearance can even vary on different parts of a roll, from one side of a roll to another.

Because variations are so common, matching sheets of reflective film should become a routine production procedure. Good design and production planning can minimize color-shift problems. Joining sections of a multi-panel graphic along a hard, dark line in the design can camouflage any change in color from one panel to an adjacent panel.

Design tips

The bigger, the better. A large mass of reflective sheeting increases nighttime visibility and improves the advertising message total impact.

Choose your colors carefully. The measure of a reflective color's luminescence is called its candlepower rating. This rating corresponds to the amount of light that is reflected back to the source. These values can vary greatly, depending on the angle at which the light source strikes reflective markings and the viewer's angle. For this reason, product specifications usually indicate entrance and observation angles when listing ratings.

Descriptions and "type" classifications for retro-reflective sheeting are listed later in this article. The chart (Table 1) compares the candlepower ratings of various colors for different classifications of reflective sheeting.

Because candlepower varies greatly between colors, the selection process is extremely important for the design of reflective graphics. For instance, orange, yellow and gold usually have comparatively high reflective brilliance, while blue, green and red have very low candlepower ratings. To improve their effectiveness, colors with low ratings should cover an area large enough to allow the motorist to see the vehicle from a reasonably safe distance.

Contrast improves readability. Studies prove that color combinations with the most contrast are more legible. Black or brown copy on a yellow, white, orange or gold background is a visually effective combination for reflective markings. Light colors on a dark background provide contrast, but with severely limited reflective intensity.

Try creating contrast — along with a more distinctive look — by using drop shadows.

Table 1: Comparative Candlepower Rating for Various Types of Reflective Sheeting
(At an observation angle of 0.1° to 0.2° and an entrance angle of -4°)

Color	Type I	Type II	Type III	Type IV	Type V
White	70	140	300	400	2,000
Yellow	50	100	200	270	1,300
Orange	25	60	120	160	800
Green	9	30	54	56	360
Red	14	30	54	56	360
Blue	4	10	24	32	160
Brown	1	5	14	12	N/A

Note: Candlepower values for reflective sheeting can vary from one manufacturer to another. The values listed in the above table represent only one manufacturer's products. To obtain product specifications on other reflective sheeting products, call your sign supply distributor or the manufacturer. The angle formed between the light source's path and a line perpendicular to the surface of the reflective sheeting is the entrance angle. The observation angle measures the angle between the line of the observer's sight and the path of the light source.

To calculate candlepower values for reflective sheeting, the amount of light reflected from a test sample is measured by light detection equipment. The reflection is compared to the amount of light emitted from a light source that simulates the light from a car's headlights, while a light detector replicates a motorist's vision.

Researchers conduct these tests at entrance angles and observation angles specified in the American Standard for Testing and Measurement's (ASTM) E-810 test procedure. Material specifications typically list test results for entrance angles of 4° and 30°, and observation angles of .2° and 0.5°. The entrance angle refers to the angle created between the line of light from a car's headlights and an imaginary line perpendicular to the surface of the reflective sheeting. "Observation angle" describes the angle between the line of light from the light source to the reflective sheeting and the line of light from the sheeting to the observer's eye.

Entrance and observation angles are important because they approximate "real-life" viewing conditions. Rarely is the light from your car's headlights and your line of sight precisely perpendicular to the reflective graphic's surface. Test results at specified entrance angles are also used to evaluate the "angularity" of reflective films, which is the material's ability to reflect light at various angles.

Some testing is conducted in controlled laboratory environments. A few reflective film manufacturers and the Federal Highway Administration (FHWA) use test tunnels to measure candlepower values. Typically, these test tunnels are approximately 50 ft. long and painted black to absorb any extraneous light that could skew test results. More commonly, researchers conduct tests using hand-held retroreflectometers, which allow them to conduct onsite measurements of a film's reflective power.

Shading letters with an opaque color provides readability when light colors are used for both the copy and the background. Drop shadows also effectively improve daytime appearance and legibility. When using light colors for reflective striping, border the stripe with a dark color so the stripe shows up during the day.

Reflective markings on tractor-trailers and large buses like this not only breathe life into the vehicle's appearance, they also protect them and other motorists from accidents.

Copy considerations.

Although there are thousands of available typestyles, keep it simple — big, bold block letters read better. Ideally, lettering should be at least 10 in. high. Spacing between letters should be approximately twice the width of the letter's stroke.

Although dark copy on a white background is usually effective, thin letters spaced too tightly together on a highly reflective background create an "overglow" — the background's reflective brilliance bleeds around the edges of the lettering, making it illegible.

Make contoured cuts. You can compensate for poor reflective intensity by printing colors with low values on white reflective sheeting. Trim the sheet on the outside of the printed graphic so that a 3⁄8-1⁄2-in. white border remains on the perimeter; this greatly improves the marking's reflective brilliance, readability and overall visual impact. Shaping the graphic in such a manner is also more appealing than a square print. Contour cutting can remedy the overglow problem.

Nesting minimizes material waste. Compared to opaque cast and calendered vinyl, reflective sheeting is costly. Many fleet graphics producers nest lettering by arranging them on a reflective sheet to maximize material usage. This helps the user obtain significant savings and a competitive edge.

Use reflective art boards. Producing and presenting a reflective art board of your client's design can generate the excitement necessary to close the sale. When you show a miniature version of reflective truck graphics, have your prospect view the design in the dark with a flashlight at eye level. Also inspect mock-ups of reflective graphics to reveal design flaws.

Nighttime safety

When we examine the basics of reflective markings, nighttime traffic safety should be the primary objective. Nearly 70% of all fatalities occur at night. People wrongly assume that reflective markings on the rear of a vehicle provide all the necessary nighttime identification.

The fact is, nearly half of all car and truck collisions are "T-bones." Moreover, these types of crashes account for more fatalities and greater property damage. Nearly all of these collisions result in some injury to the car's occupant. (Jayne Mansfield's death comes to mind.)

Imagine a large truck slowly turning in a poorly lit intersection and a speeding motorist

Standard Descriptions and Classifications for Retro-Reflective Sheeting

Based on reflective intensity and durability, reflective film products fall into six basic categories

Description	ASTM* Classification	Product Classification and Typical Candlepower Rating for White Reflective Sheeting
Utility grade	N/A	Used for short-term promotional projects and stickers. Candlepower rating: 30 to 69
Medium intensity (engineering grade)	Type I (enclosed-lens, glass-bead sheeting)	Used for computer sign cutting and screenprinted fleet identification markings. Candlepower rating: 70 to 139
Medium-high intensity (super engineering grade)	Type II (enclosed-lens, glass-bead sheeting)	Used for highway and intensity construction signage Candlepower rating: 140 to 249
High intensity	Type III (encapsulated, glass bead, retro-reflective material)	Used for highway and construction signage. Candlepower rating: 250 to 399
High intensity	Type IV (unmetallized, microprismatic retro-reflective element material)	Used for highway and construction signage. Candlepower rating: 250 to 399
Super-high intensity	Type V (metallized, microprismatic retro-reflective element material)	Used for conspicuity striping and vehicle safety markings. Candlepower rating: 700 to 1,300

*ASTM is an acronym for the American Society for Testing and Materials.

approaching the intersection. Drivers need ample warning, and side reflective markings that decorate the vehicle's full length help to alert them effectively.

National Highway Transportation Safety Administration (NHTSA) and the Federal Highway Administration (FHWA) regulations require any carriers engaged in interstate commerce to use either red-and-white retro-reflective striping (which reflects light back to its source) or reflex reflectors on the sides and rear of tractor-trailer units. These regulations cover semis, flatbed trailers and tankers with a gross vehicle weight over 10,000 lbs. The rules also cover any tractor weighing more than 2,032 lbs.

Simply put, nearly every fleet vehicle that travels anywhere must have this striping, commonly referred to as "conspicuity markings." Red-and-white markings must cover at least half the length of the sides of trailers and tractors, and must also be applied to the underside protector or bumper on the rear of the trailer. Tractors must also have reflective material applied to the bracketing system for mud flaps.

High-intensity reflective striping is an important part of any fleet-graphics program. These stripes tell the motorist the unit's length and help determine the mass and shape of the unit. By using highly reflective colors and positioning markings at a height where headlights show the greatest reflective brilliance, motorists can see the vehicle more quickly. This advance warning gives the motorist enough time to recognize the impending danger and react appropriately.

According to NHTSA research, reflective conspicuity markings provide drivers with the needed warning to avoid accidents. Government research estimates that nearly 8,000 accidents are prevented annually in the United States due to these markings. Avoiding these accidents

Harbor Graphics (Benton Harbor, MI) manufactured these reflective graphics for Jevic — one of the nation's larger common carriers — using 3M's™ fleet-marking films.

prevents $5 million in property damage losses every year. Further, the NHTSA estimates that conspicuity markings have reduced rear collisions by 25%, and side-impact collisions by 15%.

Reflectivity basics

When I started in the business, reflective sheets worked like mirrors. Appropriately, these products were called "mirror-reflective." However, mirror-reflective film's biggest shortcoming allowed motorists to see only reflective light when the headlights were perpendicular to the vehicle surface. At any other angle, the light reflected in the opposite direction, into the oblivion of the night.

Current reflective sheeting features the unique characteristic of retroreflection. Two types of retro-reflective sheeting are available: enclosed-lens, glass-bead sheeting, and micprismatic — or "cube-corner" — retro-reflective material.

"Glass bead" sheeting features a complex construction of many laminated layers. Thousands of microscopic glass beads are embedded per square inch in these layers. Sandwiched between the adhesive and the bead layer, a metalization layer is closely molded to the contoured backside of the beads, and acts as a reflector. Light passes through the film's top layers and strikes this layer. Bouncing off the metalization layer, light returns through the beads back to the light source.

Rather than glass beads, microprismatic reflective material uses an embossed geometric pattern on the sheeting's interior surface to refract the light beam. By bouncing the light off different planes of the pattern, the light is redirected back to its origin.

Various reflective film types and colors are available from manufacturers such as Arlon, Avery Dennison and 3M™. Years ago, the color selection for reflective sheeting was limited to approximately 10 colors. Today's designers can choose colors like new gold, tomato red, burgundy, light blue, purple and Kelly green.

Your selections should match the physical characteristics and budget of the intended application. Consider the substrate to which the markings will be applied.

Some reflective products can only be applied to flat surfaces. Stiff sheeting used for conspicuity striping won't conform to rivets. When applying these types of products, the film must be cut around the rivet head; special cutting tools have been developed for this application. Using the proper installation tools and techniques, softer reflective films conform to rivets, corrugations and other irregular surfaces.

Chapter 16
Metallized Vinyl Films

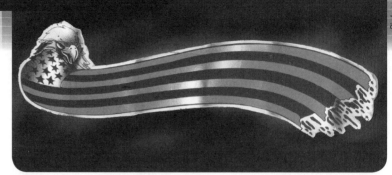

Metallized, "special effect" films have long been popular for decorating racecars, billboards and short-term, promotional signage. Some signmakers, though, shy

Metallized films aren't just for racecars anymore. Today's products are more flexible, and contain colorants and UV inhibitors that improve outdoor durability.

away from specialty films, because older products often lacked long-term durability and were difficult to cut and apply.

The good news is, newer films feature several remarkable improvements that merit a second look. Better colorants and UV inhibitors extend the outdoor life of more films. Improved release liners greatly minimize tunneling during plotter cutting, and more forgiving adhesives make repositioning during graphic application much easier.

This chapter covers metallized-film selections, how these films are made, and how you might use these eye-catching films in future signage projects.

Incredible illusions

I recently watched a woman repeatedly stroke her hand over the smooth surface of a metallized specialty film that appeared to be a rugged hunk of metal diamond plate. The amazement on her face revealed her disbelief. The 3-D illusions, which specialty films create, can certainly make one ponder how these graphics were fabricated.

The actual manufacturing process, however, is rather commonplace. Basic production steps — compounding, color matching, extruding, metallizing and slitting — are similar to those used to make other films. And, like any other adhesive-coated material, metallized films comprise sandwiched layers: a plastic film, a very thin metal layer, adhesive and a siliconized release liner.

Choices

Several plastic-film varieties, including acrylics, polycarbonates and polypropylenes, are metallized for various everyday applications, ranging from plastic potato-chip bags and wrapping paper to solar window film. In the sign industry, two of the most common metallized films are polyester and vinyl. Each film is a different type of plastic, with unique physical properties, performance characteristics and applications.

Vinyl, acrylic, polycarbonate, polyethylene and polypropylene are thermoplastics, solids that can be heated and extruded into a film, and embossed with an engraved roller.

Metallized vinyl is quickly becoming a popular polyester alternative. Its softness makes vinyl easier to plotter- and thermal-die cut, creating a more receptive printing substrate for thermal-transfer printing; airbrushing; vinyl inks; or solvent-based inkjet inks.

In contrast, polyesters are thermosets, or plastics that start as liquids and cure with heat. Once they're cured, they can't be reheated, reformed or thermal-die cut. Compared to vinyl, polyester is harder, more durable and can resist chemical spillage.

Polyester's toughness is both its strength and weakness because the facestock's hardness makes it more difficult to plotter-cut than vinyl. Still, metallized polyester films are offered in most sign-supply catalogs. These films are well-suited and cost-effective for labels, nameplates and short-term signage that must simulate the look and texture of metal.

Extruding and embossing
With metallized vinyl or similar thermoplastic films, the colorant is an integral film component that extends the product's fade resistance. Because the color's in the film, it's protected from hazards encountered with harsh cleaning chemicals, abrading car-wash brushes and day-to-day driving. In contrast, polyesters are surface-dyed, which subjects the colorant to everyday wear and tear and the sun's bleaching effects.

Prior to extruding vinyl films, the colorant is blended with the

Metallized-Film Tips

• Some specialty films are printable; some are not. Shop carefully before selecting a film to use as a print medium. You also need to follow the manufacturer's fabrication recommendations. Before going into production, test; don't guess.

• If you decide to airbrush a specialty film, make sure it's compatible with the paint system. Polyesters typically require a special topcoat, whereas vinyl doesn't. Before you get started, wipe down the film to remove any oils or other contaminants. That's also good advice before printing onto these films. Between coats, use a hair dryer to dry the paint. After airbrushing, protect your work by spraying on a clearcoat.

• If printed graphics will be subjected to gasoline spillage or abrasions, protect them with an overlaminate.

• After completing your application, always seal the edges to protect the perimeter. Clearcoating the entire graphic provides added insurance.

• When all else fails, read the product information bulletin and application instructions provided by the film manufacturer. These should cover appropriate application surfaces, substrate preparation and the minimum application temperature.

resin in the extruder's hopper, to evenly disperse the ingredients. This creates color consistency throughout the roll's entire web. The mixture is then melted and extruded through a slotted die, using a process similar to extruding acrylic or polycarbonate sheet.

When the film is hot and malleable, the vinyl's second surface — or underside — is micro-embossed, or "coined," creating a textured film pattern. Prior to adhesive coating, you can actually feel the embossing on the film's underside. The top surface, though, is smooth as glass, which is critical for printing without imperfections.

To create the micro-embossed texture, a pattern is etched into a flat, metal mold or shim. This flat plate is then wrapped around a cylinder, which serves as the stamping die.

Hot embossing creates a deeper, better impression. However, not all films are hot embossed. Polyesters, for instance, are cold embossed after the film has already been metallized. This technology provides an acrylic coating on polyester films, with the embossing within the coating.

Patterns fall into one of two classes. The first class comprises smooth, textured, metallized films. Some relatively simple patterns duplicate the appearance of brushed or leaf metals. The second class of patterns — diffraction films — is micro-embossed with fine-line prisms or images that create 3-D illusions or scatter light into a rainbow of colors.

Using a Gerber Edge thermal transfer system, the truck graphic pictured was printed white on a silver sequins pattern. Although only a streak of the special effects film was showing, the effect was attention-getting.

Silver Diamond plate silver was used in the production of the Swisher Sweets counter mat. The screen-printed film was overlaminated with a protective mar-resistant vinyl and then laminated to a non-slip pad of foam rubber.

Film pretreatment

To improve the metal layer's bond to the plastic facestock, the films typically require pretreatment, such as corona treatment, prior to metallization.

During corona treatment, an electrical discharge oxidizes the film's surface, making it rougher and giving the metallization layer something to "bite." Corona treatment increases the plastic's surface energy, making it easier for the aluminum to wet out and form a uniform layer of metal and allowing it to better adhere to the film.

A good bond between the metallized and adhesive layers becomes important during application when repositioning the graphic material. A good bond prevents adhesive/metal and metal/film delamination.

Vacuum metallization

After the film's second surface is micro-embossed, the pattern is coated with a thin layer of metal, which acts like a mirror to reflect light and create special illusionary effects. A frequently used metallizing process is called vacuum metallization. It's not the only way to metallize, but it's the preferred method to coat a uniform, thin layer of metal.

Vacuum metallizing involves evaporating a metal, such as aluminum or gold, in a vacuum chamber, so the metal vapor condenses on the film. Although several different metals are used for metallizing, approximately 90% of plastic films are coated with aluminum.

Before the metallizing process begins, the chamber must be pumped out to create a vacuum. Remaining gases will collide with the metal vapor and inhibit the metallization process. Under normal atmospheric conditions, the molten aluminum would only form a pool at the bottom of the chamber.

The aluminum turns into a vapor when it touches a heating element called a boat or crucible. Its temperature becomes extremely hot, approximately 3,000° F. As the metal vaporizes, a coil of aluminum wire continually feeds into the vacuum chamber to replace the metal that has evaporated. To maintain consistent metal thickness, the rate at which the wire is fed, the chamber's pressure, and the heating element's temperature are tightly controlled.

The cooling drum is positioned directly above the vaporization area. As vacuum metallization begins, the roll of film is fed into the chamber, where it travels around the main cooling drum. The formed vapor rises straight up to the drum, where it condenses on the film's textured second surface. The evaporation and condensation processes mimic the hot steam from your shower that fogs up a cooler bathroom mirror.

The amount of metal deposited on the film depends on several variables: the aluminum's temperature in the melting process, the speed of the film as it travels through the metallization chamber, and the atmospheric pressure within the chamber.

A chrome red metallized film was used as a decorative, eye-catching accent for the window displays of Armani Exchange during the 2004 holiday season.

The aluminum, which condenses on the film's second surface, is ultra-thin (roughly four-millionths of an inch). The thin layer acts as a reflector to create visual effects. Because the metallization layer is so thin, it's susceptible to abrasion, oxidation and corrosion. This layer is sandwiched between the transparent plastic film and adhesive coating, thus protecting it. Unsupported material — a film without adhesive — usually needs a protective coating over the metallized surface.

Best uses

Many sign people are fascinated by the look of specialty films, but they don't know how to use them. Applications range far beyond good racecar numerals.

Less is sometimes more with metallized films, because a small amount can go a long way. When used with opaque vinyl films, airbrushing, and handlettering, specialty films are excellent for lettering and logos.

Last week I watched a friend create some interesting effects by airbrushing transparent paints onto silvered, hammered-leaf vinyl. For those interested, my friend was using the Createx Auto Air Color line of automotive paint. For more information, go to www.createx-colors.com.

Drop shadows and white contours can provide contrast between lettering and backgrounds. If you're using an older film construction with rubber-based adhesive, try overlaying the film atop a cast vinyl.

This is especially important when decorating a banner. The cast film will act as a barrier and protect the rubber-based adhesive from the banner material's plasticizers. Using the cast film as a base layer can also aid in the metallized-film removal.

Dark-colored specialty films and certain textured, metallized films can produce eye-catching backgrounds for signage or tradeshow graphics. Try using specialty materials for automotive striping or decorative sign borders. Finally, consider some of the smooth or textured silver patterns for print media.

Chapter 17
Phosphorescent Vinyl

A signmaker recently told me that his company receives occasional requests for phosphorescent signage to satisfy the town's fire code. Rather than just fulfill the requested order, this creative salesman creates additional opportunities for other safety signage, such as signs for fire extinguishers and floor markings.

In this chapter, I'll review such sales opportunities, explain what makes these films glow in the dark, and explain what to look for when buying phosphorescent products.

Additional phosphorescent-vinyl uses include simply wrapping a fire extinguisher with glow-in-the-dark film to ensure that, when the lights go out, it can easily be found. Another idea is to create a phosphorescent sign that says "fire extinguisher" with an arrow pointing to it.

Everybody should know that, during a fire, building occupants should crawl along the floor to avoid the smoke. During a fire, rising smoke can obscure exit signs above door-ways. Thus, phosphorescent floor markings that mark an evacuation route can be a lifesaver.

One creative signmaker applied phosphorescent footprints to floor tile and marked a safety route for building occupants to follow during an emergency. Another solution is to outline the side of a dark hallway with photoluminescent striping — the material can be applied to a tile floor or baseboard.

To prevent people from tripping, you can apply phosphorescent strips to stairway treads, or you can apply the film just above or on a handrail. If a safety route leads up a stairway, apply glow-in-the-dark material to the stair risers. Furthermore, you can apply photoluminescent strips around a door's entrance.

Hotels provide additional sales opportunities. For example, when applied to outlet covers, glow-in-the-dark vinyl can help hotel guests locate light switches. You can also suggest a phosphorescent-vinyl evacuation schematic for each room.

If people are escaping a building from a lighted area, their eyes may need a few minutes to adapt to the darkness so they can see the phosphorescent signage.

Phosphorescent signs complement electric signage, so, when electricity fails, the glow-in-the-dark markings can direct building occupants to safety.

Compared to electric safety signage, photoluminescent vinyls have several advantages. First, glow-in-the-dark films are less expensive than electric exit and directional signage. They're also reliable. As long as the photoluminescent material has been charged, it will work. Such signage is also easy to install — you don't need to bend conduit, pull wire or be a certified, union electrician — and it doesn't consume electricity. This makes installations to older buildings simple and cost effective. Finally, phosphorescent signs don't burn out or require maintenance.

What makes the film glow?
Most pigments work by reflecting a part of the visible light that strikes them. What isn't reflected is absorbed and given off as heat. Fluorescent and phosphorescent pigments work differently — they absorb both visible blue wavelengths of light and invisible UV light and convert that energy into visible-light wavelengths.

If the light source doesn't produce these wavelengths, the photoluminescent pigment won't get a charge out of it. Only certain types of light produce these wavelengths. These light sources include daylight, tungsten-filament lightbulbs and fluorescent lighting. Other types of light, such as mercury or sodium-vapor, don't produce these wavelengths, and therefore, they won't activate a photoluminescent pigment.

Daylight and fluorescent lighting can activate photoluminescent materials using zinc sulfide in less than five minutes. Exposing these films to incandescent light can take longer to fully charge the materials, because this lighting emits more yellow than blue light.

Don't confuse fluorescent and phosphorescent materials. The two types of films are similar, but not the same. Both materials absorb UV light, which isn't visible to the human eye, and emit visible light. However, fluorescent materials only illuminate when exposed to UV light. Once the lights are out, they stop emitting. Phosphorescent materials, on the other hand, continue to emit light after the lights are turned out in the form of an afterglow.

Here's how phosphorescent material works. Electrons, which revolve around the nucleus of the pigment molecule, absorb photons from the light source. As a result, the molecules' electrons become excited or energized, boosting the electrons from their standard orbit to a higher orbit. An excited electron

Screenprinting Tips

Screenprinting with phosphorescent inks can never match the brilliance of a phosphorescent film. However, screenprinting is a great option if you have a large order for a novelty decal. Here are some ways to improve the afterglow's brilliance and duration.

- Printing the ink on a white background increases the glow's intensity.

- Diluting the ink with an extender diminishes the intensity and duration.

- Heavier ink deposits can improve the light emission's brightness. Printing with a coarser mesh (80 to 100 threads per inch) will also allow the larger pigment particles to pass though the screen and be deposited on the print surface. Large pigment particles emit more light than small particles.

- Because heat can damage photoluminescent pigments, dry the inks at the lowest possible curing temperatures. Air drying the ink is probably the safest bet.

- Because pigments can darken when exposed to moisture, clearcoat the printed phosphorescent ink.

needs heat to drop down to its original standard orbit, or it will stay in the higher orbit.

As the electron returns to its standard orbit, the excited electron undergoes a transitional state, which releases the energy it has absorbed in the form of visible light. The material glows as long as the electrons are in this transitional state. Phosphorescent pigment can be recharged repeatedly and continue to emit an afterglow for as many as 20 years.

Phosphorescent materials aren't the only substances that produce an afterglow. For example, radioactive materials also emit an afterglow as they decay. Don't worry about phosphorescent films and inks — they aren't radioactive, and they don't have an electrical charge.

Selecting the right material
Before selecting photoluminescent-vinyl film for a job, understand the job's requirements. Photoluminescent materials are evaluated according to international test standards. The granddaddy of these standards was the U.S. military specification MIL-389 1B.

Since then, more widely accepted standards have included the Deutsche Norm (the German norm), or DIN 67510, the International Maritime Organization resolution A.752 (18) and the Underwriters Laboratories UL 1994. The American Society for Testing and Materials also has published three specifications (E2072, E2073 and E2030) that cover photoluminescent safety markings.

The different standards establish specifications for the photoluminescent materials' initial brightness, as well as the materials' required duration and brightness after the initial period.

Regardless of the phosphorescent pigment in a glow-in-the-dark film, all the materials follow a two-phase curve to generate light. During the first phase, the phosphorescent material emits a strong, initial glow. After the initial period, the material can lose much of its brightness. Zinc-sulfide pigment, for example, can lose 80% of its initial brilliance within the first 30 minutes.

This phase's decay parallels the human eye's natural ability to adapt to darkness. Phase two, although less brilliant, is visible after the human eye adjusts to the darkness. In the darkness, the eyes' pupils widen to allow more light. Your eyes also become more sensitive to light. After one minute, your eyes are approximately six times more sensitive than they are in daylight. Over time, this sensitivity continues to increase. Because phase two decays at a slower and gradual rate, your eyes can see it for many hours.

Photoluminescent-material testing is both subjective and objective. The objective tests use scientific instruments to measure amounts of light. In subjective tests, researchers rely on people to tell them when they can or can't see a glowing object.

If a particular job requires certain standards, investigate whether your phosphorescent film complies. Photoluminescent-material manufacturers often list the standards their product satisfies. If such information isn't in the specs, ask your sign-supply distributor or the film manufacturer for a certificate of compliance.

Although scientific equipment best evaluates these materials, you can conduct a few simple tests in your shop. Take a few material samples, cut to the same size, and lay them approximately 6 in. apart on the floor. Then expose the samples to light for approximately 10 to 15 minutes.

Fluorescent lighting produces more UV light than incandescent lighting. Turn out the lights and observe the results for 30 minutes. Note and compare the samples' brightnesses.

Because novelty-grade materials will lose all their glow within the test period, they're unacceptable for safety-signage applications. Lower-grade engineering films, which become greyish as their brilliance diminishes, are also unsuitable for safety signage.

After testing the films for initial brightness, leave the room. An hour later, return to the darkened room and allow your eyes to adjust to the darkness (give yourself approximately five minutes). Then see if the vinyl markings are still glowing in the dark. Phosphorescent films, which are acceptable for safety applications, will be visible to the darkness-adjusted eye for at least eight hours.

Photoluminescent markings must glow for an extended period because people need considerable time to evacuate a building.

Pigments

Photoluminescent-vinyl films are manufactured numerous ways. Perhaps the most effective construction comprises multiple layers. A clear-vinyl top layer covers and protects a layer of phosphorescent pigment. Some clear-vinyl films have a UV inhibitor, which protects the pigment from sunlight in outdoor applications and prevents it from turning grey and losing any of its capacity to store energy and emit light. The pigment layer is backed by a white or metallized silver film, which reflects the pigment's glow and increases its intensity. Pressure-sensitive constructions also have a layer of adhesive, that is backed with a release liner.

The intensity and duration of a phosphorescent film's afterglow depends on the amount of pigment, the amount of light the films absorb, and the ability of the human eye to adapt to darkness.

Phosphorescent films' performance also depends on the type of pigment used. Up until approximately five years ago, zinc sulfide was the only pigment available. Today, for high-performance applications, strontium-aluminate pigments reportedly glow in the dark for 20 to 40 hours and have a sustained afterglow that's 10 times brighter than that of zinc sulfide.

Zinc sulfide stores energy very quickly. In less than two minutes, this material is fully charged and will generate a visible glow for up to eight hours. Zinc sulfide's initial glow is actually brighter than the more expensive strontium aluminate. That bright glow, however, is short lived. Within 30 minutes, the brilliance decreases to 20% of the initial glow. It also discharges energy in the form of light very quickly compared to strontium aluminate.

Strontium aluminate's brilliance of light doesn't decay as dramatically. It's visible to the night-adjusted human eye for 15 to 20 hours or longer. In the light, you can't distinguish a zinc-sulfide film from one comprising strontium aluminate. Both look yellowish-green. In the dark, their appearances change. Zinc-sulfide films give off a yellowish light. Strontium aluminate has a greenish glow.

Some municipalities specify a grade of phosphorescent film. However, for most applications, vinyls using zinc-sulfide pigments should suffice.

Chapter 18
Inkjet Media

Who needs topcoated, inkjet vinyl these days? With the advent of low-cost, solvent printers that can print on most vinyls, sales of topcoated vinyl have declined, and some predict their obsolesence.

Think again. Just when the market assumes a product is doomed, someone develops an innovative, rejuvenating twist.

I'll discuss topcoated-vinyl upgrades and how they work, and suggest how to select inkjet-printing components.

When choosing media, the user must consider whether it suits the shop's printing technology or the installation's environment. Ferrari Color (Salt Lake City) produced this mounted print for Bill Graham Presents, a West Coast concert promoter.

A delicate balance

Material selection would be much easier if every inkjet vinyl worked with every printer. But, life doesn't work that way. Interrelated printing components, such as the printer, vinyl, ink, and clearcoat or overlaminate, create a very complex chemical mix. And, because the technology continually evolves, many long-term, real-time, test results aren't available for all components.

In an ideal world, purchasing decisions would be a simple matter of price comparison. However, that also isn't how it works. In fact, inkjet-vinyl selection is especially difficult, because photorealistic reproduction, color gamut and outdoor durability rely on the inter-relationships of the printer, ink, vinyl and (when used) overlaminate. Any equation factor that changes can alter a delicate balance.

Top-coated inkjet vinyl

Your vinyl and printer must match. If you're printing with a thermal-inkjet system that uses only waterbased inks, you need a vinyl with a special topcoating or receptor coating so the ink will adhere to vinyl. Without the coating, waterbased inks bead up on the film's surface, just as water beads up on a newly waxed car.

Coatings can do more than bind an ink to the substrate. Some enhance the ink's appearance by controlling its flow. Although the receptor coating absorbs the ink, it also needs to maintain the dot structure on the media. This prevents ink spreading, which destroys any print definition. If the dots bleed together, the printed image can appear fuzzy and darker than intended. Good dot integrity creates crisper images and intense, vibrant colors.

Some complex, topcoating systems comprise two layers. In this type of system, the ink passes through the top layer, and the basecoat absorbs it. The top layer acts as a barrier and protects the ink system from UV light, water, chemicals and abrasion. And, the coating must accept heavy ink saturation.

Some topcoated vinyls reportedly work with dye-based and pigmented inks. Prints produced by these ink types may look significantly different.

Pigmented inks comprise large molecules many times the size of dye molecules. Because of their size, pigmented inks struggle to penetrate a topcoat and tend to remain on the surface. Conversely, smaller dye molecules seep readily into the coating.

In addition to making an ink adhere well to the vinyl substrate, topcoats also accelerate ink drying. Other coatings form a protective barrier between the ink and the vinyl that provides dye-based inks with some water resistance.

Although UV-inhibiting, topcoat additives provide some fade resistance, dye-based ink isn't outdoor-durable. A printed image's durability, however, largely depends on the ink, not the coating or vinyl. Although a coating provides limited ink protection, clearcoats or overlaminates may still be needed.

Chemistry differentiates topcoated, inkjet vinyls. Companies that specialize in coatings for different types of plastic films have developed hundreds of varieties. Each coating caters to a specific printing technology and, therefore, has unique, physical properties.

When evaluating products, remember that all topcoated vinyls aren't created equal. Chemistry, coating thickness and film consistency affect print quality, durability and print-production waste. A great price means nothing if a shop loses 10% of its work.

Some new inkjet films' topcoats must be cured using an infrared heater. They're reportedly compatible with dye-based and pigmented inks. When these materials are heated, topcoating encapsulates the colorant, and both fuse to the vinyl through a polymerization process.

The process' high heat won't distort the printed image. Encapsulating the ink inside the film protects the printed image from moisture, chemicals and abrasion. Clearcoating and overlaminates aren't needed, but the system requires some additional equipment.

Total system approach

Some vinyl companies use a total-system approach that integrates components in a warranted graphics solution that combines the printer, inkjet vinyl, ink, overlaminates, surface-protection masking and printing profiles. For example, 3M® has formed alliances with printer manufacturers Roland DGA Corporation, Hewlett-Packard and VUTEk.

Selecting the pre-package system approach eliminates most of your guesswork and also minimizes printing catastrophes. The systems approach is clearly the safest choice, because manufacturers have done the costly and time-consuming lab and field testing. In this package deal, the signmaker gets peace of mind should a problem occur. However, the years of R&D and the warranty security come with a pricetag.

Mix and match

Although some OEMs warn of the potential component-substitution problems, most sign-makers mix and match when selecting consumables. If that's your approach, be sure to "test, don't guess." Testing and evaluating media, inks, clearcoats and overlaminates don't need to be complicated or require expensive lab equipment. But, it will take time.

Keep your testing simple. The most reliable durability test entails simply placing a printed sample in sunlight to see how long it lasts without fading. The results of printing different inks on the same vinyl, or one ink on different vinyls, may vary greatly. To most reliably evaluate image and color reproduction, print on different materials, and select the best.

Because vinyl stretches, the topcoating must stretch with it. After printing and clearcoating a test sample, pull and see what happens to the printed image. If everything works, the image will stretch too. If it doesn't, the ink and topcoat will crack.

For sign applications, fade resistance is also critical. The best test of outdoor durability is outdoor exposure. You'll know a graphics system will last two years in the sun only if you test these exact conditions and times. This is much more reliable than accelerated weathering of a test sample in a weatherometer.

Outdoor graphics, especially vehicle graphics, endure such demanding conditions as moisture, chemical spillage and ink adhesion. Ink adherence to vinyl is critical for these applications.

Because solvent-based inks incorporate different solvents, they won't adhere equally well to uncoated vinyl. So, test ink-to-vinyl adhesion using two screenprinting tests that gauge ink adhesion. The first is called the thumbnail test — just rub your thumbnail over the test print and see if ink scratches off.

The second is the cross-hatch test. Lightly score the ink surface with an X-acto® knife. Then, apply transparent tape to the ink and pull it off perpendicular to the film's surface.

Consider these factors when evaluating inkjet vinyl:
• Contrast the color's richness between prints.
• Check for fuzzy edges or detail loss.
• Note how long it takes for a print to dry.
• Consider the film's handling characteristics during vinyl application.
• Determine if the release liner impacts print processing.

Overlaminates and clearcoats

Any sign that's subjected to a demanding environment should be clearcoated or protected with an overlaminate. That covers most outdoor applications. When making product selections, pair calendered and cast vinyls to matching overlaminates.

Although inkjet inks' fade resistance has improved, any printed image needs protection from abrasion and cleaning chemicals. Vinyl overlaminates are available with different finishes, such as matte, luster and satin, that complement the print's appearance.

Overlaminates also provide vinyl graphics with added stiffness, which can ease film handling during graphics installation. However, the print still needs surface-protection masking. With or without an overlaminate, premasking a print can prevent vinyl from stretching during film application and repositioning. The mask also protects the print surface from squeegee scratches.

High-tack application tape is usually too aggressive for digital prints. During one application, high-tack tape pulled the overlaminate, ink and the topcoat off the vinyl substrate as I removed the tape. Although much of the problem could be attributed to a topcoat failure, we used the wrong tape. For digital prints, ask your distributor for an ultra-low-tack, surface-protection masking.

Liquid laminates, which are cost-effective alternatives to overlaminates, may be either spraypainted or applied with a roller. Be sure to use the right one for your ink system. Some solvent-based clearcoats are designed to work with either solvent-based or water-based inks. Waterbased clearcoats are generally compatible with solvent-based inks, but may not be suitable for some water-based inks. Before using any clear coat with any ink system, the best advice is to "test, don't guess.   For any product questions, ask your distributor.

Before production, test all components to ensure the vinyl, ink and clearcoat are compatible. Testing can simply entail coating a print and checking to see if any adverse reaction occurs, or verifying that you achieve good ink-to-vinyl adhesion.

Conclusion

Many people choose inkjet media during printer evaluation. For equipment and consumables consideration, your sign-supply distributor should provide the necessary technical information.

Vinyl and printer manufacturers can also provide good advice for selecting appropriate products. Once you've determined a successful materials combination, stay with it.

Chapter 19
Overlaminates

Photo courtesy of R Tape

Years ago, the limited range of vinyl overlaminating films simplified product selection. Used on everything from tanker markings to safety labels, a few polyester laminates were all that was available. But today, the range of products includes a variety of facestocks, adhesive systems and release liners.

Common facestocks include polyester, polypropylene, polyvinyl fluoride (PVF), polycarbonates, and cast and calendered vinyl. Although glossy overlaminates sharpen color contrast, matte finishes soften colors and contrast, thus hiding printing imperfections. When used for indoor applications, matte finishes also minimize the glare of fluorescent store lighting.

The two types of overlaminate adhesive systems that exist are heat-activated and pressure-sensitive. Though heat-activated films are cheaper and protect paper prints, pressure-sensitive laminates are the only choice to use with vinyl.

Pressure-sensitive overlaminate material, as shown here, provides the requisite strength and durability for protecting safety labels.

Pressure-sensitive overlaminates protect photographs and digital prints, as well as P-O-P and tradeshow graphics. Compared with heat-activated overlaminating films, pressure-sensitive films provide greater durability, optical clarity, ease of use, and diverse finishes and textures. Although cold laminates cost more initally, this is offset by less frequent print damage, lower scrap cost and less reworking.

Because the heat of thermal overlaminates can damage thermal-transfer, piezo, and phase-change inkjet prints, pressure-sensitive overlaminating films are the only option. Heat-activated overlaminates can also damage or distort some plastic print media. Even some types of paper media will warp at elevated temperatures.

Heat-activated adhesives still offer lower cost and better adhesion to paper prints. The bond of a cold, glue adhesive to a print isn't so stable. Pressure-sensitive adhesives are coated onto the release liner as a liquid, and, after the coating process, the adhesive is dried in a curing oven and then laminated to the facestock. (This process is called "transfer coating", as opposed to "direct coating".)

Still, the adhesive maintains a degree of fluidity. A fluid adhesive can cause the overlaminate to slip or shift on print media, especially if the two materials expand and contract at different rates. When this shifting occurs, a "tunnel" can develop after the print is rolled.

Carefully consider the liner used with pressure-sensitive films because it can affect the finished print's appearance. Using a paper liner, some cold glue overlaminates exhibit a mottled, or "orange peel," pattern, which is especially noticeable over dark colors. The mottling effect results from the adhesive picking up the mirror image of the rough, paper release liner surface. To correct the effect, manufacturers are improving the paper's smoothness. On the other hand, overlaminates with polyester liners, which are perfectly smooth, aren't prone to mottling.

Key considerations

Options facilitate selecting the right product for a specific application, but too many choices can complicate a purchasing decision. Picking the best overlaminate requires matching product performance characteristics to the application's demands.

The overlaminate, print media and printing method must be collectively selected as part of a total system solution. Also, duly consider the project's environment. Durability depends heavily not only on environment, but also upon the printing process and raw-material interactions with each other. As always, the customer's expectation is very important.

A process of elimination streamlines options. Here are some important considerations:

- What is the application? It's important to pick the right material for the job. Whether the task involves floor graphics, perforated window graphics or fleet markings will dictate the appropriate overlaminate.
- What substrate will receive the graphics? Will it be smooth, such as flooring, an acrylic or polycarbonate sign face, or will the graphics be applied over rivets or corrugations?
- What environment will the graphic endure? Will there be temperature extremes, UV light, water or pollution? Will it be subject to acidic or alkaline cleaning agents, or could it be exposed to vandalism, grease or road salts?
- Under what lighting conditions will the graphic be viewed?
- What are the job's durability requirements? What are the customer's expectations?
- Is the overlaminating film compatible with other components of the graphics system?

Printed fleet graphics

Most fleet graphics don't need overlaminate protection. However, exceptions include inkjet or electrostatic graphics. Even screenprinted vinyl and graphics printed via thermal transfer can require overlaminates in special cases.

Overlaminates protect printed vinyl from scratching, UV degradation, abrasion, moisture, dirt, grafitti and chemicals. Be cautious of ex-tended-durability claims, such as assertions that a laminate will double the print's outdoor life.

Durability really depends on the ink system and its exposure to UV light. Consider the installation site's latitude, altitude and amount of time exposed to direct light. Exposure to severe UV light can bleach out most colors. The right overlaminate can provide a certain amount of added protection, but don't expect miracles. All fluorescent colors — and even certain non-fluorescent colors, such as magenta — will always be vulnerable.

Some, but not all, overlaminates provide UV protection by either absorbing or reflecting UV light. With one type of overlaminate, substances in either the facestock or adhesive absorb UV light. The absorbers convert UV radiation to heat, which prevents print damage. The film's UV blockage is cumulative; eventually, the substances reach a saturation point at which the film can't absorb any more light. The other type of overlaminate merely reflects UV rays.

When choosing an overlaminate for screen-printing applications, test the components before using them in production. Because they need a thicker layer of ink, screenprinted graphics typically require an overlaminate with a heavier coating weight of adhesive. An overlaminate with a thin adhesive coating will bridge the edge, causing a slight, but notice-able, air pocket.

The substrate and durability requirements impact the appropriate selection of vinyl film and overlaminate for fleet graphics. Two key considerations are the type of substrate and durability requirements. The rule of thumb dictates using a calendered-vinyl overlaminate with a calendered vinyl film and a cast-vinyl overlaminate with a cast vinyl film (Fig. 1).

Similar films typically expand and contract at the same rate. If not, the overlaminate can delaminate from the base film or substrate. A tunnel could also form between the two films. In my opinion, the cast-vinyl option is usually the safest bet for fleet applications, especially for graphics applied over rivets and corrugations.

To make installations easier, use a cast film with a repositionable adhesive. More forgiving adhesive systems help prevent "pre-adhesion" of the graphic to the substrate. In turn, the installer needs to reposition the graphic much less frequently. This is important because excessive handling can cause delamination.

Overlaminates can ease vinyl application, because the added thickness not only makes the graphic more rigid and easier to handle, but also simplifies the graphics' removal. Some experienced decal installers can even apply these stiffer prints without application tape.

Large, one-piece digital and screen printed graphics, however, should be covered with premium-grade, low-tack application tape after lamination. The application tape protects a graphic against scratches during installation and prevents graphic distortion. This is espe-cially important when applying graphics in hot weather, when vinyl can stretch like a piece of warm taffy.

I learned the importance of protecting overlaminates on bus markings more than 20 years ago when I sold a bus program to a city in central Illinois. The graphics were printed and clearcoated. However, within three months, the nylon bristle brushes used daily in the washing system had abraded protective clearcoats and inks right down to the base vinyl. The complaint went to court, and the plaintiffs won. The lesson of this story: Printed bus markings always need overlaminate protection.

Public vehicles, such as buses, are common targets of marauding vandals armed with cans of spraypaint. Specially coated PVC overlaminates can protect vinyl graphics from this urban artistry, but for extremely demanding applications, expensive polyvinyl fluoride (PVF) overlaminate films provide the utmost protection (Table 1).

Because PVF is a low-energy plastic, paint has difficulty sticking to it. These films, which also provide excellent UV and abrasion resistance, are ideal for graffiti-proof markings, but because these films don't stretch, they can be used only for flat applications.

When fabricating bus graphics, overlaminate protection is essential to prevent clearcoat and ink abrasion.

Proper cleaning is critical in preserving PVF's anti-graffiti properties. Clean dirty graphics with a nylon-bristle brush and citrus-based cleaner, such as Simple Green®. Solvents aren't recommended because they can erode or etch the surface, which causes paint and dirt to stick to it.

Perforated window graphics

Perforated window-graphic films used for fleet identification should always be protected by an overlaminate, preferably cast-vinyl films. In addition to protecting the graphics from abrasive cleaning systems, overlaminates prevent dirt and water from collecting in the film's holes, thus causing the adhesive system to fail. In all cases, the graphic's outside edges should be sealed using a commercial edge sealer.

Tankers and cement-trucks

Increasingly popular, clearcoating and liquid laminates provide great short-term protection for promotional pieces, but they don't outperform overlaminates. Graphics on chemical tankers are frequently subjected to chemical spillage. Acids, caustics and solvents can quickly erode the ink and clearcoating system of printed fleet markings, as well as leech the plasticizer from pigmented vinyl, embrittling it and causing the film to crack. Cement-truck graphics, which are easily damaged by caustic cement and harsh cleaning chemicals, should also be overlaminate-protected.

Although vinyl overlaminates are usually best for vinyl markings, chemical tankers and cement trucks are an exception. For protecting markings subject to chemical damage, polyester overlaminating film was the best choice 20 years ago and remains so today. Polyester exhibits outstanding chemical and temperature resistance — features that also make it an excellent protective film for warning labels (Table 2). But, because polyester isn't conformable, it can't be used over rivets, corrugation or compound curves. Furthermore, it shouldn't be used to protect such flexible surfaces as awnings.

Table 1: Characteristics of PVF Overlaminates

- Excellent for demanding applications
- Excellent outdoor durability
- Graffiti resistant
- Excellent UV protection
- Excellent abrasion resistance
- Very expensive

Table 2: Characteristics of Polyester Overlaminates

- High strength
- Excellent clarity
- Withstand high temperatures
- Excellent chemical resistance
- Outdoor durable
- Good abrasion resistance
- Rigid, inflexible

Table 3: Characteristics of Polycarbonate Overlaminates

- Excellent for floor graphics, in-store graphics and exhibit graphics
- Withstand abrasion and rough handling
- When used for floor graphics, provide excellent slip resistance
- When used for in-store graphics, velvet texture eliminates glare from overhead lights
- Thick, rigid covering with excellent lay-flat characteristics
- Very expensive

Graphics laminated with polyester also require special handling. To prevent tunneling, delamination or bubbles between the overlaminate and base film, graphics are best stored and shipped flat. During installation, take special care to avert delamination. Furthermore, to prevent edge lifting, graphics applied to the surfaces of tankers that haul caustic or petroleum-based products should be coated with a commercial edge sealer.

Table 4: Characteristics of Polypropylene Overlaminates

- Moderate temperature resistance
- Good clarity
- Good chemical resistance
- Good durability
- Can be used outdoors
- Scratch easily

Floor graphics

Because floor graphics must withstand extraordinary abuse from pounding foot traffic, dirt, grease, grit and chemical cleaners, several types of overlaminates are designed for these applications.

Polycarbonate makes the best choice for floor applications because this tough, resilient film can withstand daily traffic, and its matte texture prevents slippage. However, polycarbonate film is expensive (Table 3). Thus, calendered vinyl overlaminate is an economical, but less durable, alternative. Whichever film you choose, all floor graphics should be waxed with a product recommended by the film manufacturer to prevent edge lifting.

To ensure that a floor graphic is slip-resistant, overlaminates are rigorously tested. The industry standard is American Standard for Testing Materials (ASTM) D2047. The ASTM test checks the overlaminates's coefficient of friction, the force required to move one material over another. In essence, the test measures the traction a pedestrian could expect as he/she walks on a floor graphic.

Liability is a major concern for retailers, graphics producers and raw-material manufacturers. If a shopper falls on a slippery graphic, someone could get sued. Of course, the biggest litigation targets are companies with the deepest pockets. For this reason, some manufacturers provide an insurance policy with floor-graphic materials, which protects screenprinters, signshops and retailers from litigation ensuing from an accident.

In-store, P-O-P and exhibits

As digital printing has grown in the sign industry, so has the use of overlaminates for signage, interior-wall and trade-show graphics, and P-O-P applications.

Thick polycarbonate overlaminates are also often used for the protection of exhibit materials. This rigid protection offers excellent lay-flat characteristics, and its velvety texture eliminates any glare from overhead lighting. Polycarbonate can also withstand the rough handling encountered in trade-show environments.

Vinyl overlaminates are offered in various finishes.

Because of their resistance to the wear and tear of dirt and chemical and heavy pedestrian traffic, polycarbonates are recommended for floor graphics.

These films cost more than polyester but are easier to laminate and less prone to bubbles and edge lifting. A cost-effective alternative to polycarbonate or vinyl, polypropylene overlaminating films now include a variety of 2- and 4-mil facestocks in several finishes. For display graphics, 5- and 10-mil polypropylene films are also available (Table 4).

Chapter 20
Application Paper

A signmaker's life would be easier if one type of application paper served all purposes. At one time, only a few choices existed. Today, however, product offerings have mushroomed to accommodate various "special" applications. Because many signmakers aren't aware of these products offerings and how they differ, in this chapter, I'll discuss application-tape basics and characteristics, and answer some FAQs.

What signmakers consider "application tape" or "application paper" is often referred to as "premask" or "prespacing tape" by screenprinters. Basically, it's the same product and performs the same functions — it protects vinyl graphics in storage and handling and aids in film installation.

"Application tape," or "application paper," protects vinyl graphics in storage and handling and aids in film installation. Today, product offerings are plentiful to accommodate various "special" applications.

"Transfer tape" is also frequently used interchangeably with "application tape." However, the two terms refer to different products — transfer tape refers to a transfer adhesive, which is an adhesive coated onto a release liner.

Application tape offers more than paper and adhesive. To optimize better adhesive anchorage to the facestock or improve unwind characteristics, various coatings can be applied to the paper prior to adhesive coating. Adhesive additives are frequently used to promote adhesion to the release liner.

Application tape is available in a few paper-thickness grades and several adhesive-tack levels. Standard tapes, which measure 4 to 4.5 mils thick, weigh and cost less.

Most application-tape products sold within the sign industry comprise standard-weight paper, which features a medium- or high-tack adhesive. For most small sign projects, these do the trick. However, in the large-format, fleet-graphics market, heavyweight, premium-grade tapes, which measure 5 to 5.5 mils thick, are preferred.

Most manufacturers offer several adhesive-tack levels for each paper grade. The most popular tack levels and applica-

Table 1: Tack Level and Applications

Tack Level	Applications
Low and Ultra Low	Digital prints, surface protection for plastic, glass and metal
Medium	Large- to medium-sized graphics
High	General sign work
Slightly Higher Tack	Difficult-to-transfer vinyl films
Ultra High	Thermal-die cuts

tions are listed in Table 1. For small, computer-cut lettering or general sign work, use high-tack tapes; for large- to medium-sized lettering and banner applications, use a medium-tack tape; finally, for large-format digital prints, use low tack tapes. Selecting the right tack requires testing. Your distributor or tape manufacturer can provide samples for your evaluation.

For most small sign projects, standard-weight paper, which features a medium- or high-tack adhesive, is sufficient. However, for large-format, fleet-graphics applications, premium-grade tapes measuring 5 to 5.5 mils thick are preferred.

For a more thorough tape evaluation, consider the following test, suggested by a tech service manager for a vinyl manufacturer: To test vinyl sheets, laminate various tapes. Then store the sheets for a month and try to apply the vinyl. You should be able to easily remove the application paper from the applied vinyl — without tearing or leaving behind adhesive residue.

Avoid mishaps

Consider the following helpful tips when working with application tape:

- Whether you laminate application tape via hand or a laminator, avoid trapping air bubbles between the vinyl graphics and application tape. Bubbles and wrinkles in the application paper can reappear in the applied graphic. If you inspect a graphic's adhesive side — after having removed the release liner — you often can see wrinkles and bubbles forming.
- During the lamination process, avoid stretching the application tape — stretched tape usually shrinks and causes the vinyl graphic to curl.
- If you can afford one, invest in a laminator. A properly set-up laminator can apply application tape to graphics with minimal wrinkles and bubbles. If you can't afford a laminator, consider purchasing an application-tape dispensing system.

The WEBERmade (Carlos, MN) dispensing system bolts onto your workbench with a couple of clamps. It also includes a clutch, which allows you to adjust the roll's unwind tension.

The Mask-Rite tape system comprises two rollers through which the application tape feeds. Simply roll the application tape over your graphics.

- Use a single tape sheet (rather than overlapping pieces) to cover the graphic. Otherwise, a fine line of tiny air bubbles will appear where the tape pieces overlap.
- For screenprinted and digitally printed decals, thoroughly "cure" inks and clearcoats before applying the tape. Solvents in uncured inks and coatings often cause the decal and tape to adhere to each other, which hinders tape removal once the graphic is applied. Plus, solvents can cause ink to delaminate from vinyl during tape removal.

For screenprinted or digitally printed graphics, curing the ink system is critical. The rule of thumb is to wait 24 hours — allowing inks to thoroughly dry — before clearcoating or overlaminating. Then wait 24 hours before applying the application tape and trimming the print.

- After a vinyl graphic has been "taped," it should be applied shortly thereafter. Prolonged storage can increase the bond between the tape and vinyl, making tape removal difficult following application. This condition worsens if the graphic is stored at elevated temperatures.
- Wet applications generally require more time for the vinyl's adhesive to bond to the substrate. Thus, you need to allow for more time before removing the tape. It's best to apply the graphic dry. However, if you must use application fluid, use the least amount possible.
- If an application-paper tape isn't releasing easily following an application, lightly spray the paper with application fluid, wait approximately 30 seconds and then remove it. The application fluid will penetrate the paper facestock and soften the tape's water-based latex adhesive, causing it to release more easily from the graphic.
- When removing the application tape, carefully pull the tape 180° against itself (see above). This tape-removal procedure prevents you from pulling the graphic off the substrate. To minimize edge lifting and prevent the squeegee from scratching the graphic, use a squeegee covered with a low-friction sleeve to re-squeegee the entire graphic, especially the edges.

Storage rules of thumb

Application-tape manufacturers package their products in plastic sleeves, then box them in corrugated containers to protect the tape from light, dirt and humidity. If you're not going to use the tape immediately, keep it in its box. Tape not stored in boxes is exposed to light, which can cause yellowing. Even shop lighting can yellow application tape.

This yellowing occurs because the light begins to degrade the natural rubber adhesive. Although this degradation isn't generally significant to adversely affect product performance, it's aesthetically unpleasing.

Yellowing caused by light detracts from an application tape's appearance. However, storing a pressure-sensitive product in an excessively hot environment is more harmful. Prolonged exposure to temperatures above 85°F prematurely ages an adhesive and degrades its performance. Years ago, I met one sign maker who kept his application tape in the refrigerator. Wonders never cease! My advice is to store your beer in the fridge, and keep your tape in your shop's coolest area.

As tapes age, their performance

To prevent pulling graphics off a substrate, carefully pull the application tape 180° against itself. Depicted here is Avery Dennison Graphic Div.'s A6 opaque series film.

often declines and becomes problematic. Old application tape can "block-up" on the roll — the adhesive binds to the facestock's first surface, making the product difficult, and often impossible, to unwind. All pressure-sensitive products have warranted shelf lives; for application tape and premask, it's six months. Consequently, you should rotate your tape inventory. The material first entered into inventory should be used first. Furthermore, handle your application tape with care — damage to a roll's ends impairs unwinding the tape, and he paper easily tears where it's damaged.

Finally, application-tape rolls should be stacked upright. Stacking them horizontally causes a flat spot on the tape roll, making unwinding more difficult.

Premium-grade tapes

At tradeshows, signmakers frequently ask me why they should invest in premium-grade application tape when standard-grade tape accomplishes the same task at a lower cost. The answer is simple: Premium-grade tapes help eliminate headaches. They minimize errors and material waste, and they help you produce a visually appealing finished product in less time.

Not surprisingly, you'll pay more for premium tape, but a good product is worth a few pennies more per square foot. Here's why:

- Thick, premium tapes facilitate vinyl applications; thin and cheap tape tears more easily when it's removed following an application. This can slow installation tremendously.
- Because premium tapes are thicker, they laminate to vinyl graphics more easily, while reducing wrinkles and bubbles. Their thickness gives more body to flimsier films and results in a smoother vinyl with fewer wrinkles in the squeegee process.
- When installing a project outdoors, a premium application tape can be a lifesaver. During strong winds, heavier transfer tapes are less likely to tear. Plus, fewer mishaps in the shop and on the job site equate to less material wasted (vinyl and transfer tape) and less time spent on remake work.

Best of all, after you finish squeegeeing a graphic, the premium tapes can be pulled off quickly in one sheet, not in bits and pieces.

- While laminating application tape to vinyl, the tape is sometimes stretched. The resulting tension generates a force that causes both the tape to return to its original shape and the vinyl graphics to curl. Often, the problem can be avoided if extra care is taken when laminating the transfer tape to vinyl.

Curling is less likely to occur when using a premium-grade application tape, because thicker tapes are harder to stretch — less stretch equals less curl.

- Premium tapes work better than standard grade tapes in humid environments. Extreme humidity causes various vinyl-graphics problems. As the tape's moisture content increases, the paper "grows." Extreme humidity can also cause stretching and curling.
- Application tape is a paper product. Wet paper can disintegrate. And the longer it stays wet, the more likely it is to fall apart. Because heavyweight paper has more substance than standard-weight paper, it holds up better during wet applications.
- Regardless of how careful you are when working with transfer tape, accidents occur. However, you can minimize your tape-lamination lamentations by switching to heavier, premium-grade tape. Thicker tape is less likely to wrinkle during the lamination process.

Roll Slitting Tricks of the Trade

For years, the sales people of R Tape Corporation have explained the advantages of our "factory cut" rolls of application tape. In slitting material, R Tape utilizes very expensive and sophisticated industrial machinery that unwinds, slits and rewinds the rolls of tape. This heavy-duty machinery is ideal for producing rolls without imperfections.

Many of our competitors, by comparison, utilize lever-type slitters, which are commonly referred to in the sign and screen print industries as "bologna slitters". While it's possible to cut rolls of tape with acceptable edges using this lever-type of equipment, often "bologna" cut material exhibits imperfections that give sign makers fits. A common problem with "bologna" cut application tape is a nick in the paper, caused by a blade that isn't sharp or that has nicks. Rolls with nicked edges frequently tear. Another problem is a roll with a dented or crushed edge. Dented edges frequently result because one side of the cutting blade has a beveled edge, which tends to slightly crush the edge of the roll. A third problem is gapping between the layers of the tape. Gapping often results when too much pressure is exerted by the operator in the cutting operation. Excessive force compresses and crushes the roll, causing gaps that expose the adhesive on the application tape to air. This allows it to dry out and, consequently, loose tack. Finally, in using the lever-type of equipment, glue balls often form on the edge of the roll because the heat generated from the friction of the blade melts the latex adhesive and fuses it together. Nicks in the paper, crushed edges, and glue balls can cause the application tape to snag and tear as it is unwound. When the application tape tears, it often falls onto and ruins the vinyl lettering.

While the case in favor of "factory cut" rolls is very strong, we realize that many R Tape distributors have chosen to slit their application tape. To help our distributors do a better job of roll slitting, we've compiled a few tips. At first glance, the operation of a "bologna" slitter seems relatively uncomplicated. But, as with many things in life, it's always looks a lot easier than it really is. In fact, mastering the roll slitting operation involves learning the idiosyncrasies of the equipment along with a few tricks of the trade. Cutting some material can also require modifications in blade type and cutting procedures. For example, cutting polypropylene tape can be more demanding than slitting paper tapes or polyethylene films.

1. The slitting blade. Having the proper blade and keeping it sharp may not be all that you need for great roll slitting, but it certainly helps in getting the job done right. Always use a sharp blade when cutting either paper application tape or application film. Regular sharpening should be part of any preventative maintenance schedule. Dull blades can crush the edges of paper tapes, as well as distort or shatter a film facestock. Believe it or not, a blade can be too sharp. (This usually isn't something to worry about when slitting rolls of application paper, but it can be a problem when cutting rolls of film.) An over sharpened blade can produce what is called "angel hair" (fine strings of film) when slitting a film. An overly sharp blade can also result in a concave cut. Overly sharp blades can even cause the adhesive to smear on the cut edge of the roll. Smeared adhesive causes the tape to stick together, resulting in difficult unwinding. In contrast, a dull blade frequently causes a convex edge. As a regular practice, you should check the evenness of your slit rolls by placing a straight edge across the roll's edge and visually inspecting the cut.

Selecting the right blade for the job is also critical. Unfortunately, there isn't a general purpose blade that is suitable for all applications. Three basic configurations are as follows: (a) a single bevel blade with no back bevel; (b) a single bevel blade with a slight back bevel; and (c) a double bevel blade with both bevels at equal angles. The bevel angle of the blade is very important in producing a straight cut. Incorrect bevel angles can result in either convex or concave roll edges.

Use a single bevel blade in cutting application paper. A blade with a slight back bevel should be used in slitting narrow width rolls of film, such as 1/2", 1" or 2". Rolls exhibiting a concave condition result from blades that are too sharp. Rolls with a convex edge are frequently caused by a blade that is too dull. Angel hair indicates a blade that is too sharp. Use a double bevel or wedged blade when cutting larger rolls of film, such as 12" or 24".

Cutting paper application tapes requires a blade that is flat on one side and beveled on the opposite side. The beveled side is ground to a 45º angle. Remember that the side of the blade with the bevel will slightly crush the edge of the application paper. With a single-bevel blade, it is possible to cut application film between two and six inches wide. Single bevel blades can also be used to trim the edges of a roll of film. In cutting rolls of film between six and 24 inches wide, however, a double bevel blade must be used. The double bevel blade acts as a wedge that pries the roll apart. In cutting application film, friction generates heat between the side of the blade and the edge of the film. The extreme heat generated in the process can not only melt and fuse the adhesive together, but it can also embrittle the tape to the extent that the film can fracture as the roll is unwound.

To stabilize a blade during the cutting process, some people have mounted a six inch plate to the outside of the blade. The purpose of the plate is to minimize blade vibration. By doing this, the blade makes a smoother, more even cut. The plate also serves a secondary function of kicking the off cut end of the tape away from the roll and along the shaft.

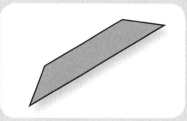

Fig. 1: Use a single bevel blade, such as this, in cutting application paper.

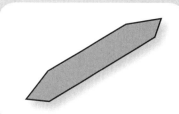

Fig. 2: A blade with a slight back bevel should be used in slitting narrow width rolls of film, such as 1/2, 1 or 2 in. Rolls exhibiting a concave condition, result from blades that are too sharp. Rolls with a convex edge are frequently caused by a blade that is too dull. Angel hair indicated a blade that is too sharp.

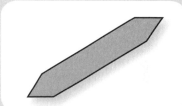

Fig. 3: Use a double bevel or wedge blade when cutting larger rolls of film, such as 12 or 24 in.

Hard plastic films, such as polypropylenes, may require thinner blades made of extremely hard steel. Although double bevel blades were primarily designed for cutting film, they are also great for slitting narrow widths of paper tapes (2" and smaller). This type of knife produces a straighter cut.

Lever-type or "bologna" slitters are inexpensive pieces of equipment that do an acceptable job for many applications. This equipment, however, has its share of limitations. Whether a operator cuts paper or film tapes, the maximum length of roll that can be cut is approximately 300 feet.

2. Wet or Dry Slitting. Although application papers and films can be successfully cut with or without a fluid-cooled blade, a cool blade usually produces better results. An uncooled blade results in the latex adhesive melting and producing "glue balls".

3. Dry cutting. Slitting without a coolant can be accomplished, but it requires a sharp blade and the right cutting technique. In cutting either paper or film application tapes, use the lightest pressure to accomplish the task. The weight of the operator's hand is usually all the pressure that is required. As my daddy always told me, "You don't need to muscle the tool. Let the tool perform the task that it was designed to do." Some operators mistakenly try to speed up cutting by pulling excessively on the blade handle. Excessive pressure crushes the edges of the application; it can cause gapping; and it can cause the blade to overheat.

Selecting the proper cutting speed is a matter of trial and error, experience and sometimes a little luck. Longer rolls require a slower cutting speed. Tighter rolls and harder facestocks should also be cut more slowly. One rule of thumb is to start at a slower speed, such as 100 rpm, and then increase the speed as required. Don't cut so slowly, however, that you produce inferior cuts. (NOTE: Different brands of slitters have different features. Not all slitters have variable speed control. Some operate at one speed only.)

In cutting application paper, both the chuck and the blade should be engaged. Chuck speeds for some slitters are typically 400-450 rpm, with the blade speeds set at 200-215 rpm. (NOTE: The blades on some types of slitters are free-wheeling.) Different materials may cut better at different chuck and blade speeds. Often, in cutting some hard films, better results are achieved by disengaging the blade and allowing it to rotate freely. (NOTE: With some slitting equipment, the blade may not disengage completely.) Running the blade while cutting hard films, such as polypropylene, will cause excessive friction. This friction causes the blade to heat up, which can cause the adhesive to melt and ball up. Some operators will lubricate the blade with a silicone spray to reduce friction and heating.

In cutting hard films, some operators will run their equipment at a slower speed, such as 250-275 rpm. As the blade cuts though the film, they will gradually increase the chuck speed to as high as 380 rpm. This slow acceleration generally produces a smoother cut. Keep in mind that as you begin to cut into a log of application tape, the surface speed at the outside of the roll is much greater than it is at the core. That's because the outside of the log travels a greater distance per shaft revolution than it does closer to the shaft. As the blade cuts more deeply through the tape, the surface speed gradually decreases. Increasing the chuck or mandrel speed compensates for the change in surface speed in the cutting process.

4. Wet cutting. Cooling the blade with fluid usually results in smoothly cut edges with no glue balls. Slitters can usually be retrofitted with a misting system for less than $500. The best cooling systems utilize two nozzles, spraying both sides of the blade. The nozzles should be directed at the tip of the blade. Typically, the recommended air pressure is 40 lbs. The amount of misting solution is regulated by a volume control knob. In operating the system, always monitor the level of the fluid in the tank. Running the tank dry requires the pump to be primed.

Commercial mist cooling fluids should be used for cutting operations. If fluid is used to cool the blade, some precautions should be taken. For the best results, use the least amount of fluid to cool the blade. Cutting film requires much less fluid than cutting application paper. Excessive fluid can potentially deaden and contaminate the adhesive. Solutions of water and alcohol may soften the adhesive, discoloring it and turning it a pasty white. This typically isn't a problem. If the tape is dried adequately, the discoloration usually disappears. After slitting rolls of tape, dry them by placing them on vertical rods or on corrugated board. The corrugated board allows the moisture to wick away from the edge of the tape.

Chapter 21
Vinyl-Banner Basics

Many vinyl-film companies spend significant time testing and evaluating their products' suitability with banner substrates. Because banner substrates include polyester, nylon, canvas and flexible signface material, there's much to test.

With myriad products on the market, researchers can't test every one. Thus, this chapter will focus on vinyl banners. I'll discuss banner-material selection, as well as vinyl-application techniques, and banner painting, installation, storage and cleaning options.

Selecting banner material

A vinyl banner substrate is usually the best choice when applying vinyl graphics. Such banner materials typically

Selecting the right banner material is critical. Therefore, you should consider, and test, a material's flexibility, opacity, gloss, smoothness and thickness prior to making a selection. When applying vinyl graphics, a vinyl banner substrate is usually the best choice.

comprise a polyester scrim embedded in white vinyl. These vinyl materials are cast or extruded, and are manufactured similarly to cast and calendered films. During casting, the scrim is coated with layers of liquid vinyl; during extrusion, layers of hot PVC bond to the polyester scrim.

Selecting the right material is critical. Therefore, rely on your sign-supply distributor, and study the manufacturer's technical bulletins. Once you find a good combination of materials, stick with it.

When evaluating a banner material for cut-vinyl graphics, screenprinting or digital printing, vinyl and banner manufacturers consider such characteristics as flexibility, opacity, gloss, surface smoothness and material thickness. Test methods that check such performance properties are essentially straightforward. In fact, you can perform these in your shop prior to selecting a product.

Flexibility. Although banners shouldn't be folded, check a material's flexibility. Fold the banner and check whether the substrate returns to its original shape. Because some customers will store banners folded — even after you've instructed them otherwise — you don't want the banners to show creases.

Opacity. If you decorate the banner on both sides, make sure the material has enough hiding power to prevent one side's graphics from showing through the opposite side.

Technical bulletins often document opacity as a percentage, such as 96% or 100%.

Gloss. High-gloss material can obscure applied or printed graphics. For printed graphics, a matte finish is usually preferred. When reviewing product specifications, look for materials with a gloss level under 10, at a viewing angle between 45° and 60°.

Smoothness. If you're going to print onto the banner, rather than decorate with cut-vinyl graphics, examine the surface's smoothness. Smoother surfaces print better, whether you're screenprinting or digital printing.

Material thickness. A thinner banner substrate, such as a 10-oz. material, is easier to process through a printer. For outdoor banners, customers expect durability. Thus, persuade them to use a high-quality, heavy-duty banner substrate, such as 13-oz. or 15-oz. material.

Heavier fabric is stronger, less likely to tear and allows for wind slits. Material weighing 15 oz. or more is generally designed for decoration on both sides.

Furthermore, heavier outdoor material generally has fewer, but thicker, threads per inch. Banner substrates with thicker threads aren't as smooth as indoor substrates, whose fabric features a higher thread count. Materials with more threads per inch have a smoother surface, but, as the surface becomes smoother, it loses strength.

Dyne level. Whether you're applying cut graphics to a banner surface or printing on it, dyne level measures a substrate's surface energy. Inks and pressure-sensitive films stick less easily to banner materials with low-surface energy, because the ink, or adhesive, won't readily wet out the surface.

A minimum dyne level of 36 is desired. Banner material with a higher dyne level allows the adhesive on vinyl films and inks to more easily wet out the surface for good adhesion. Compared to other materials, PVC banner material's advantage is a consistent dyne level.

Hue. Whites can vary. Vinyl banner material comes in two different hues: bluish white and yellowish white. For outdoor advertising and digital applications, bluish-white material is generally preferred. On the other hand, screenprinters often select yellowish-white material, because it costs less. If you're decorating the banner's front and back, check the material's whiteness on both sides, so the banner's front looks the same as its back.

Many signmakers are tempted to strip off graphics from old banners and recycle the material. However, if you value your time, which is probably worth more than $50/hour, this doesn't make any sense. The cost of banner material is lower than your labor cost to clean an old banner.

Ask the right questions

When you initially review a project with a customer, ask the right questions. This way, you can determine which material weight and type to use. Here are some questions to consider:

- What's the banner's intended purpose?
- Where will the banner be used?
- Will the banner be mounted flat against a building, or mounted between two poles?
- What city ordinances apply to banners?
- How long does the customer want the banner to last?
- What's the customer's budget?

"Most customers who walk into my shop don't have a clue about what type of banner they want," said Butch Anton of Superfrog Signs & Graphics (Moorhead, MN). He continued, "As a sign professional, it's my job to steer them in the right direction — advising them about size, color and content — so I can best satisfy their business needs."

Anton primarily produces stock-size banners. He keeps nearly six of each popular size, such as 3 × 6, 3 × 8, 3 × 10, 4 × 8 and 4 × 10-ft., in inventory at all times. He only inventories white banners, because he can easily paint them to change the background color.

Before using a banner substrate, store your raw materials as you would other signage materials: out of direct sunlight and in a temperature- and humidity-controlled area.

Keep banner designs simple

Good banner designs are simple — don't use eight words, when four will do. Don't use an overly ornate typeface, when a simple one is more legible. Furthermore, limit your number of elements, and emphasize a primary message, such as a store special. Also, allow enough white space, or open area, to create an uncluttered look and improve readability.

Bigger banners are better. Sell your customer on the idea that bigger banners are more noticeable, more readable and generate more store traffic. Bigger banners also mean higher revenues for your shop.

When you see an attractive banner design, photograph it for future reference when designing your own banners. A portfolio can also be a useful selling tool.

It doesn't have to be white

Generally, brighter colors attract attention. Instead of black block letters on a white background, try something different. Vinyl banners are available in various colors. Signmakers, like Anton, often paint the background color.

"Instead of selling a plain-white banner," Anton explained, "we add value to the signage by adding color. By doing so, we can charge an additional $30 for the banner." To paint the backgrounds, Anton uses Ronan Aquacote waterbased paints, which he applies using a foam roller.

He further explained, "We roll a thin coat of paint onto the vinyl banner. The paint quality is so good, it covers completely in one coat. Typically, we can paint a 4 × 8-ft. banner in 10 minutes."

By directing high-volume fans onto the painted material, the banners are dry to the touch in approximately 30 minutes and ready for vinyl applications.

After two hours, the paint is bulletproof — it's so hard you won't pull any paint when you reposition vinyl or remove application tape. However, if you tear off a little paint, you can easily touch up your work.

Ideally, you should wait a day before rolling vinyl banners. The waiting time is certainly much less if you use solvent-based paints or lettering enamels.

According to Anton, "With waterbased paints, I can paint the banner, dry it, decorate it with vinyl, and ship it within a day. This is important when dealing with customers who need their banners right away. Plus, if you can provide more than your competitors, you can charge more for your services."

Before you apply graphics, various paints can decorate vinyl banners. Although sign enamels can be used, many signmakers prefer waterbased paints, because they dry faster.

Avery Dennison's Alan Weinstein noted, "Not all paints are compatible with all vinyl banner materials. When you mix numerous raw materials together, some complex chemistry takes place. The banner substrate, paint, vinyl and clearcoat, or laminate, must all be compatible. If they're not, problems can occur."

For long-term, or demanding applications, it's best to use cast films. For most shops, price often dictates the use of cast or calendered vinyl to decorate banners. Thus, for short-term banners, many professionals believe it's unnecessary to use expensive cast film.

All banner-material surfaces should be considered contaminated; therefore, requiring proper preparation.

Cast films are intended for long-term, or demanding, applications. Not surprisingly, because price often dictates whether cast or calendered vinyl is used to decorate banners, many professionals believe it's unnecessary to use expensive cast film for short-term applications.

Although some industry professionals use a non-abrasive, soap-and-water combination, soap residue could potentially contaminate the surface and adversely affect adhesion. Because it safely removes plasticizer, isopropyl alcohol (IPA) is generally preferred for cleaning banners prior to vinyl application.

Don't use any cleaner stronger than IPA, because such strong solvents as lacquer thinner and acetone can destroy the material. Solvents that can melt PVC and be absorbed into the banner material will eventually attack the vinyl's adhesive. Furthermore, some solvents can draw out plasticizer from the banner and affect adhesion.

Plasticizer makes PVC supple and soft. Plasticizer, however, tends to migrate. Banner-substrate manufacturers try to formulate the PVC so the plasticizer fuses into the material and doesn't migrate to the surface.

Awning materials are more heavily plasticized than PVC banner material. Plasticizer can soften an adhesive and cause a film to slip on a substrate's surface. Thus, very tiny tunnels, called work tracks, can form in the vinyl.

Before applying graphics, tape the banner securely to your work surface using quality, 2-in. masking tape. Keep the banner substrate tight during the vinyl application to prevent wrinkles in the applied graphic. Whenever possible, perform a dry application.

To aid the application process, use a hinging technique, such as a side hinge. When creating a hinge, tape one side of the graphic to the banner or work surface. Then, reach under the graphic and remove enough release liner to expose a manageable amount of adhesive.

If you must apply the vinyl wet, use a commercial-grade application fluid, such as ActionTac, Splash or RapidTac. Don't use a homemade concoction. Use the least amount of fluid to accomplish the job. After removing the application tape, re-squeegee the entire graphic to prevent edge lifting.

After you remove the application tape from the vinyl graphics, always re-squeegee the entire graphic. Then, using a rivet brush and heat gun, burnish the vinyl into the banner's textured surface. The heat and brushing technique will aid the adhesive flow and result in improved, ultimate adhesion. Many signmakers who specialize in banner work prefer medium-tack application tape, because, after installing vinyl graphics, the less aggressive tape is easier to remove from the banner substrate.

If you're applying graphics to a very large banner, consider taping the banner to a wall or any other vertical surface. This way, you won't need to fight gravity as it pulls the graphic to the substrate. By working on a vertical surface, you won't have to stretch across the entire banner from one side of your worktable.

After applying vinyl to the banner material, allow the banner to lay flat for at least a day, so the vinyl graphics' adhesion builds.

Clearcoating

In most cases, clearcoating, or using an overlaminate on a short-term promotional banner decorated with vinyl, is unnecessary. On the other hand, outdoor banners printed using waterbased inks require more protection. In these cases, lamination is a requirement, not an option.

If you're producing a banner that will be used repeatedly, invest in high-quality vinyl and clearcoat the banner using waterbased Frog Juice Sun Screen Clear 7000. The clearcoat, which can be either sprayed or brushed on, protects the banner from air pollutants and the sun's bleaching effects. Clearcoats also help seal the edges to prevent edge lifting. Unlike overlaminates, they won't tunnel. Many outdoor banner materials' heavy texture won't allow an overlaminate to adhere without silvering or tunneling.

Construction and installation

Strong winds can easily damage a banner — 80-mph wind gusts can create as much as 35 lbs./sq. ft. of wind pressure. For a 4 × 8-ft. banner, this equates to more than 1,100 lbs. of force. Thus, you should consider some windproofing options.

Some signmakers believe that any banner larger than 30 sq. ft. requires wind slits to reduce wind load. Wind slits, also called wind pockets, are semi-circular, or crescent-shaped, cuts in the banner substrate. These slits, which vary from 3 to 5 in. in diameter, are spaced approximately 18 to 24 in. apart. To prevent the banner material from tearing, stitch the slits' ends.

However, not everybody agrees that wind slits are a good solution. In fact, many manufacturers won't warranty their materials if wind pockets are cut in them — the slits weaken the substrate.

According to Butch Anton, wind slits don't work. "They detract from a banner's appearance, they weaken the material, and they don't do a very good job at relieving wind pressure."

If you use a rope to install a banner, interweave the rope from one grommet to another, either along the banner's top or bottom. This method distributes the wind load over a greater area than if you just tie the banner at the corners. Keep the ropes holding the banner in place taut so the banner doesn't flap in the wind and possibly rip.

A more effective solution, bungie cords, stretch under windload and can prevent the banner from tearing. Ropes are too rigid and less forgiving than bungie cords. Furthermore, bungie cords allow banners to bounce in the wind, won't loosen or slacken and don't require re-tensioning.

Banners that stretch across a street can be attached from grommets to steel cables above and below the banner. The cables can be mounted to either buildings or poles. Ropes tied to the corners can pull the banner taut and prevent it from flapping in the breeze.

If you sew your own banners, rather than purchase pre-made ones, use a heavier material, double-stitch your hems, reinforce the corners and carefully install the finished product.

To reduce the chances of ripping, the material's edges should be folded over and double-stitched. To strengthen the hem, sew additional fabric under the banner's folded edges. If you attach grommets along the banner's edges, reinforce them by adding washers on both sides of the banner, under the grommets. Grommets should be spaced 2 ft. apart.

Cleaning, storing and shipping

If your customers want their banners to last longer, they should periodically wash them with a mild detergent and warm water. At the very least, exterior graphics should be washed once or twice a year; a banner should be cleaned before it's stored. Instruct customers not to use cleaners that contain abrasives or solvents, or a pressure washer (the sprayer's force can cause edge lifting), to clean banners.

Banners won't last forever. However, banner owners can extend their soft signs' lives by properly caring for them. Banners should never be folded; instead, roll them around a 6-in., fiberboard core with the graphics facing outward. Rolling banners with the graphics facing inward can result in edge peeling and tunneling.

If you're sending a customer a double-sided banner, ship it flat.

If you're shipping numerous vinyl banners, a used release liner placed between the individual pieces prevents abrasion.

Chapter 22
Plastic Substrates

One afternoon, I was finishing an interior-graphics installation at a convenience store when a passerby unfurled an annoying barrage of questions and comments. Feeling a little put off, I gave her a brusque reply that abruptly ended the conversation and sent her on her way. Under less-hurried circumstances, I could have politely and patiently explained what I'm going to cover in this chapter: selecting rigid plastics and applying vinyl to these surfaces.

This illuminated sign fabricated for Plassein Intl., a flexible-packaging manufacturer, features an etched top comprising Cyro Industries' (Rockaway, NJ) Acrylite® AR sheet and Oracal's (Jacksonville Beach, FL) 651 intermediate, translucent vinyl. Advision (Pittsburgh) designed and fabricated the sign.

Plastics basics

You probably have a good idea of what plastics are. Signmakers use them daily to create backlit signs, photo-mounting POP displays and yard signs. Standard materials I'll review for this article include acrylic, polycarbonate, styrene, corrugated polyethylene, formed polyvinyl chloride (PVC) sheet and fiber-reinforced plastic.

These commonplace substrates have some remarkable characteristics. Many plastics can weather extreme heat and cold, as well as the degrading effects of UV light and the stresses of gale-force winds. One might not guess that acrylic's optical clarity equals or surpasses glass, but it's true.

Plastics can be classified as "high polymers." These very complex materials comprise mega-molecules called polymers — chains of much smaller, building-block molecules called monomers.

Plastics divide into two subcategories: thermoplastics and thermosets. Thermoplastics are solid plastics that become malleable with heat. PVC, acrylonitrile butadiene styrene (ABS), polyethylene (PE), polycarbonate (PC) and acrylic are thermoplastics commonly used in the sign industry. Thermoplastics are relatively easy to fabricate and decorate; plastics manufacturers generally offer detailed instructions and suggested techniques. These book-lets address equipment recommendations for saws, drills, blades, routers and bits required for fabrication.

Conversely, thermosets are liquids that solidify when heated. Unlike a thermoplastic, once a thermoset cures, it can't be reheated and reformed. Typical thermosets include polyester, epoxy and polyurethane.

Acrylic

Acrylic remains the mainstay product for backlit sign faces. Popular brand names, such as Dupont's (Wilmington, DE) Lucite® and Atofina Chemicals (Philadelphia) Plexiglas®, are household words. Acrylic suits many applications because it's clear, resistant to chemicals and weatherable under virtually all conditions. Compared to polycarbonate and flexible-face material, it's also an economical alternative.

Manufacturing acrylics entails two common methods: casting and extrusion. To mold a plastic sheet, the cell-casting process involves pouring liquid, acrylic resin between two pieces of glass separated by a gasket. The gasket's thickness — which can have a diameter up to 5 in. — determines the cast sheet's thickness. Cell casting produces an enhanced product, but the time-consuming process adds to the sheet's cost.

A more cost-effective way to manufacture cast acrylic is continuous casting. During this process, liquid plastic is poured between twin, polished, stainless-steel belts. A spacer cord that divides the belts governs sheet thickness. The sheet is heated and cooled on the casting line to control the curing process.

The more common and cost-effective acrylic-manufacturing method is extrusion, or calendering. During extrusion, plastic pellets are heated, and the molten plastic is forced through a slotted die that governs sheet thickness. This sheet is then run between rollers, which polish the plastic's surfaces. As an extrusion variation, the "continuously manufac-tured" process involves extruding the sheet through a die, and then calendering between rollers, which output the molten sheet at the required thickness.

Don't stress out
The extrusion process imparts mechanical stress, or orientation, into the sheet. Most of this stress flows toward the "machine direction," which determines which way the shrinkage will shift.

Cell-cast acrylic also has an orientation, but it's bi-axial. This means the stress occurs across the length and width of the sheet, and shrinkage occurs evenly on the acrylic's surface.

Extruded- and cast-acrylic sheets appear to be identical. However, there's one major difference: Extruded acrylic is more sensitive to grooves and cracks, and thus more prone to breakage.

Be careful when working with acrylic and other plastics. Cutting, drilling, gluing, painting and solvent cleaners subject the sheet to myriad mechanical and chemical stresses. One stress factor may not fracture a sheet of material, but these pressures accumulate. Such stress-causing problems and materials build internally, until the plastic reaches its breaking point.

Annealing plastic sheet after fabrication can alleviate some stress-related buildup. This involves heating the sheets in a recirculating oven. A rule of thumb is to heat the sheet for one hour per each millimeter of its thickness, at 175° F. However, be certain to check the manufacturer's product bulletins for recommended time and temperature.

Stress caused by cutting a plastic sheet can be eliminated by scraping and smoothing rough edges with a notch tool, which plastics distributors offer. After scraping, pass a propane flame under its edges to polish the plastic.

Brown kraft paper or polyethylene masking typically encases both sides of a plastic sheet. When you're fabricating a plastic sheet, don't remove the masking — it will protect the sheet from hot plastic chips that can fuse to its surface.

Sometimes masking can be difficult to remove, especially if it's been atop the sheet for a long time. Some sheet manufacturers recommend removing masking by rolling it around a wooden dowel rod. Heating the sheet for 60 seconds at 350°F can expedite paper-mask removal. Soaking the paper with a solvent can also help.

If any adhesive has transferred onto the sheet after mask removal, it can usually be cleaned with a soft rag moistened with isopropyl alcohol. Removing paper masking may generate a static charge in the sheet, and static attracts dirt. To neutralize sheet static, wipe the surface with a damp, lint-free rag.

Thermoforming

Extruded acrylic's lower forming temperature makes it easier to thermoform. Lower forming temperatures translate into shorter cycle times, speeding production.

Thermoforming encompasses preheating a plastic sheet, and then forming it with a vacuum against a mold. If vinyl adorns the sheet, problems can occur if high temperatures assault the substrate for long periods of time. Vinyl manufacturers — such as 3M (St. Paul, MN), Arlon (Santa Ana, CA) and Avery Graphics (Painesville, OH), recommend temperatures less than 380°F for less than eight minutes. High heat can degrade the adhesive's performance and change the film's gloss level and color.

Thermoforming sheets without sufficient preheating can also create vinyl problems. Insufficient heating causes excessive mechanical stress for the film, resulting in cracking or discoloration.

When using vinyl graphics on a plastic sheet that will be thermoformed, apply the graphics to the surface opposite the side encountering the mold to prevent damage. When using a male mold, apply vinyl to the plastic's first surface; when using a female mold, decorate the sheet's second surface.

Vinyl applications

When fabricating backlit signs, vinyl can be applied to the sheet's first or second surface. When decorating a backlit signface, overlap seams rather than butting panels together. All

Federal Heath Sign Co. (Dallas) shows how to apply vinyl to poly-carbonate sheet. First, remove the 3 x 3-ft. sheet's protective film.

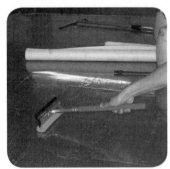
After applying a soap-and-water solution, squeegee the surface to remove dirt and other impurities.

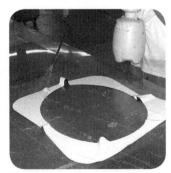
Turn over the vinyl, spray its back with soap and water, and squeegee it before applying vinyl to plastic.

films shrink; when this happens, butt joints will separate and expose a thin line of substrate.

If you are required to join multiple panels of the same color of film, ensure that the colors match to prevent a noticeable shift. At the least, make sure all films come from the same roll or lot number.

For second-surface, clear-plastic applications, cover the applied graphic with a diffuser film. The film scatters the light from the lamps, which produces an even appearance under illumination without any "hotspots."

When you overlap one film layer over another, edge bridging can occur. This is especially obvious when graphics are applied to the second surface, but bridging usually vanishes during adhesive outflow.

If one layer of vinyl film must register over another, I suggest using marks in the graphics' corners to aid alignment.

Wet applications

When you apply vinyl graphics to acrylic and other plastic sign substrates, I advise a wet application. I know this contradicts most of what I've written about application fluid, but acrylic applications are challenging because film and plastic attract each other. For such jobs, just one disastrous application made me believe in application fluid.

Being frugal, I've tried mixing my own application fluid. I added 1/2 teaspoon of Joy dishwashing liquid and 1/2 teaspoon of isopropyl alcohol to a quart of water. This concoction works in a pinch, but it's inferior to commercial-grade application fluid, which I now use whenever necessary.

However, wet applications can be problematic when installing large graphics; the fluid tends to cover everything, including the film's release liner, which must stay dry. Application fluid can cause the release liner's siliconized layer to flake off and contaminate the adhesive, which creates noticeable spots on an illuminated sign.

Center the cut-out over the sheet. Federal Heath used 3M™'s custom VT Series colors on its 3630 translucent film.

Nice buns! The first element of the Burger King graphic is added to the face. Use nylon squeegees; inspect them for nicks prior to use.

The backing film prior to removal.

Polycarbonate

Polycarbonate's (PC) advantage is its great strength. It withstands wind and abrasion much better than acrylic. Understandably, that advantage also carries a higher price tag.

PC's drawback is its tendency to yellow and degrade under UV light. However, plastics chemists have improved PC topcoatings, which has helped remedy the problem.

PCs absorb, and then outgas, moisture. Manufacturers often recommend drying PC in an air-recirculating batch oven for at least one hour prior to thermoforming or graphic application. This procedure helps remove the sheet's internal moisture. Vinyl applied over insufficiently dried PC will trap outgassed moisture, which causes bubbles.

After drying, prep the surface with water and a mild, nonabrasive cleaner. Because the PC is softer than most plastics, it's susceptible to scratching, so clean it carefully. Don't use any solvents — although PC is strong, it's susceptible to subsequent stress-induced cracking. Avoid using brushes, squeegees or other cleaning devices that could scratch the sheet's surface. After washing the surface, rinse with clean water and dry with a soft cloth to prevent water spotting.

Having prepared the surface, install graphics using commercial application fluid. Washing the surface and using application fluid after drying the sheet may not make sense, but don't worry. Neither washing nor wet application will compromise PC sheet.

Polystyrene

Polystyrene is a cost-effective option for interior retail signage and promotional displays. It's also a popular screenprinting substrate, because it's smooth and printable with various inks. Unlike polypropylene and polyester, no surface treatment, topcoating or special preparation is necessary before printing.

Styrene outgasses and summer heat can exacerbate this condition. In many cases, outgassing breathes through permeable film.

A more complex styrene formulation, polystyrene, is more stable and less prone to

After removing the backing, squeegee the elements into place.

After a final squeegeeing of the elements, a white diffuser film is applied to enliven the graphic.

The finished face. Federal Heath manufactures these vinyl-on-plastic signs for Burger King up to 12 x 12 ft. According to Raymond Nugent, the company's plastics-plant manager, such signs are often pole-mounted at restaurants.

outgassing. Nevertheless, it won't work outdoors. I investigated a vinyl-graphic failure several years ago where styrene was the substrate. The sign was mounted inside a bus-shelter enclosure with southern exposure. As heat built in the tight confines, the sign cooked and formed gigantic bubbles in the applied graphics.

There's only one reason anyone would use polystyrene outdoors: It's cheap. The cost of replacing the sign once it's abused by outdoor conditions won't make using polystyrene worth your time or money. Avoid it altogether for exterior signage.

Corrugated plastic board

Corrugated plastic boards, such as Coroplast®, usually comprise polypropylene or poly-ethylene. The most common sign-industry thicknesses range from 2 to 6 mm. As a cost-effective sign substrate, it's hard to beat. In the short term, it resists chemicals and holds up well outdoors.

Because corrugated plastic possesses low surface energy, it's corona treated. Changing the sheet's surface energy enables screenprinting inks to wet out more readily and adhere to the surface. However, the corona treatment doesn't last forever; typically, it starts to degrade after approximately two years.

Solvent-based, or UV, inks typically decorate corrugated plastics. Before printing, play it safe and test ink adhesion using a cross-hatch or thumbnail before an actual production run.

Low surface energy also inhibits vinyl adherence to sign blanks. To test a substrate for vinyl readiness, pour a small amount of water onto the surface. If the water forms a smooth film, the surface is protected; if the water forms droplets, the surface has lost its adhesion enhancement, and the graphics aren't likely to stick.

In a pinch, you can flame-treat polyethylene or polypropylene. First, clean the surface with isopropyl alcohol. Use a propane torch with a spreader tip to wave the flame over the surface. The outer blue flame should touch the plastic. Properly done, the treatment shouldn't even warm the material. After completion, you can cover the surface with vinyl.

When applying vinyl with a squeegee to corrugated sheet, a zillion tiny bubbles form readily in the flute's valleys. The hard edge of the squeegee rides along the corrugations' crests, spanning the valleys and trapping air. To prevent this, use a rivet brush, instead of a squeegee, for your applications. The brush's stiff bristles will help the vinyl conform to the grooves of the plastic surface.

Rigid PVC panels

Rigid PVC foamboards, such as Alcan Composites' (St. Louis) Sintra®, Vycom Corp.'s (Moosic, PA) Celtec® and Kommerling's (Pirmasens, Germany) Komatex®, are popular signmaking materials. Available in myriad colors, PVC sheets range from 1 to 30mm thick, with varied densities and finishes.

Using standard woodworking tools, PVC is very easy to fabricate. The material can be cut using a utility knife and table, band or panel saw. It can also be drilled, routed, heat-line bent and vacuum-formed. Thin sheets may be painted, screenprinted and decorated with vinyl.

Consult manufacturer literature regarding special cutting tips and such requirements as cleaning and surface preparation, paints and inks, blade and bit types, cutting speeds and edge-finishing suggestions. To mount photographs and digital prints onto PVC, use transfer or spray adhesives to laminate the sheet.

During manufacturing, PVC forms closed cells that inhibit water absorption. This makes it a very good outdoor-sign material. Although PVC board is very durable, care should be exercised when clamping it because too much pressure will cause dents.

An integral PVC component is a release agent, which can hamper adhesion. To promote a strong bond, wipe the surface with acetic acid — white distilled vinegar is a common source — and then quickly wipe it with toluene. Remember, inhalation of toluene and other solvents can be hazardous to your health, so follow manufacturer recommendations, work in a ventilated area and use chemical gloves.

PVC foam has a natural static charge that attracts dust, dirt, hair and plastic chips like a magnet. Briskly wiping down the sheet with a dry rag only creates more static. Instead, wash the surface with water and a mild detergent.

Most plastics' inherent disadvantage is a high expansion coefficient. I learned that term's definition the hard way. It means that a plastic sheet can expand and contract dramatically when exposed to temperature extremes. When I first worked with plastics, I didn't know about this phenomenon. On a cool, late-October day, a co-worker and I installed a very large sign comprising several, expanded-PVC sheet panels. The ponderous sign wrapped three sides of a building.

I thought I understood the manufacturer's instructions. However, I mistakenly believed that construction adhesive and finishing nails were perfectly acceptable for applying PVC to wood.

Next summer, I'd learned my lesson. When the sun warmed the panels, they expanded and buckled. We had to remove them and start over. When installing plastic-sheeting signs outdoors, the sheet must be allowed flexibility. An effective trick is installation within a track or channel. Oversized, slotted fastening holes also allow sufficient movement.

Fiber-reinforced plastic sheet

As a PVC alternative, certain hybrid sign substrates, such as American Acrylic Corp.'s (West Babylon, NY) Lumasite®, Nudo Products' (Springfield, IL) Fiber-Lite® and American Fiber Technologies' (Waterbury, CT) FiberBrite®, are quite serviceable materials.

Naturally, these sheets have strengths and drawbacks. Compared to PVC, the fiber-reinforced sheets are much stronger and less prone to edge cracking. However, this strength makes the sheets heavier and unwieldy for some projects.

During sheet production, manufacturers add fiberglass to plastic resin and cast it into shatterproof sheets. These fibers give the sheet its tremendous strength, making it very rigid and stable. Compared to other plastics, it has a low coefficient of expansion, meaning it expands and contracts more like metal than other plastics.

Furthermore, these sheets don't lose strength, even after years of outdoor exposure. Available in various thicknesses, sheet sizes and translucent or opaque colors, the sheets are compatible with standard sign paints, screenprinting inks and pressure-sensitive vinyl for decoration.

Several cements and adhesives suit these hybrid substrates. Foam tapes and mechanical fasteners can also be used for assembly. Most saw blades, router bits, drills and fabricating techniques used with other plastics can be used to process fiber-reinforced sheeting. When working with this material, be sure to wear safety goggles and a dust mask. Unlike other plastics, sawing and routing fiber-reinforced plastic produces fiberglass dust, which is a serious health hazard.

Words to the wise

Flat-surface applications should be simple, but the dreaded tiny bubbles that sometimes creep beneath a vinyl graphic still plague many signmakers. Whether you create graphics for plastic, glass or metal, or apply them wet or dry, here are a few tips to consider:

- Carefully apply application tape to vinyl graphics because wrinkles and bubbles in application tape yield wrinkles and bubbles in your applied graphics. A smooth surface provides optimal results. Furthermore, a glass-covered table is ideal. A table grooved from numerous cuts will generate bubbles. If you remove the release liner from vinyl after masking, you can often see where little bubbles have formed on the adhesive;
- Use a firm, quality nylon squeegee for your applications;
- Before use, inspect your squeegees for nicks; a nicked edge will generate bubbles. Rubbing its edge back and forth along another squeegee's center will sharpen the edge;
- Use overlapping strokes with good squeegee pressure during applications; and
- When completing wet applications that entail overlaying layers of vinyl, allow enough time for the first layer of vinyl to dry and adhere before applying the subsequent layer.

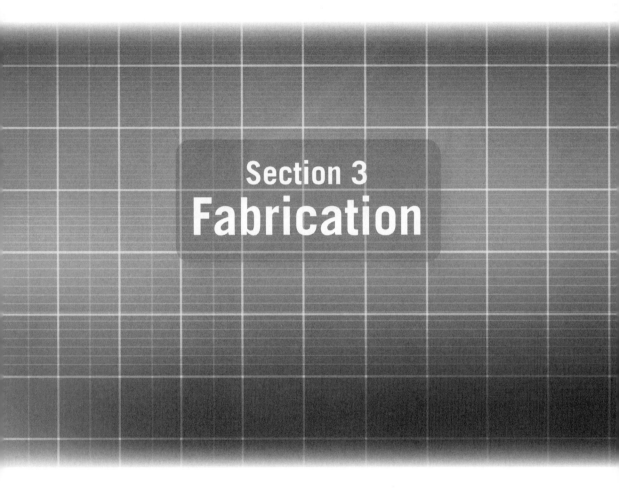

Section 3
Fabrication

Chapter 23
Design Basics

Judy Eck of FDC Graphic Films Inc. (South Bend, IN), asked me to write a chapter about how to use special-effects vinyls, such as Avery's shade-shifters or R Tape's Vinylefx metallized films. Judy believes most signmakers are fascinated with such unique films, but they might not know how to use them.

Without the white outlines and black drop shadow, the copy would be lost against the background. The boat name's special-effects vinyl gives the lettering a contrasting visual texture with the background colors.

Shortly after promising Judy that I would write about special-effects vinyl, I designed a graphic using these films. On paper, the design looked great. However, instead of making a prototype to see what the design really looked like, I charged ahead with a production order of 500 pieces.

When the finished product crossed my desk, I was mortified. The background was so sparkly and busy, I couldn't read the copy. The design failed because I failed to follow basic design rules, which is the topic of this chapter. (Sorry, Judy, I changed my mind. I'll write about special-effects films another time.).

The design appraisal

Asking questions is an important first step in the sign-design process. Gaining a better understanding of a customer's business will help you select the appropriate colors, typeface and other design elements.

Some of the questions include the following:

- How do your customers and prospects currently view your business?
- How would you like to be viewed as a company?
- What are your current advertising and marketing themes?
- What colors, logotypes, typefaces, design motifs and slogans are you currently using?
- What liberties can be taken with corporate logos and colors?

Such design elements as color selection, typeface and logotype must be in sync with the theme your client wants to convey.

For example, contrasting colors usually aid readability. The most readable color combinations include: black on yellow, yellow on black, white on black, and blue on white.

What type of typeface?

When I worked in advertising more than 30 years ago, I learned that serif type, such as Times New Roman, was more readable than sans serif, because the first words we read in elementary-school primers are in that typeface family. The morning newspaper is also set in a serif font, with which we're more familiar and more comfortable reading. "Readability," however, isn't the same as "legibility." Times New Roman may be more eye-pleasing, but this doesn't mean it's more legible.

So, what do we mean by legibility? Generally, sans-serif fonts can be read more easily from a distance than fonts with serifs — the small, embellishing strokes that finish off a letter's main strokes. However, this doesn't imply that you should always use sans-serif copy. In fact, most designers believe that serif type is more likely to be read because it's more pleasing to the eye, and they're right.

As with other things in life, size matters, especially if your sign is viewed from a consider-

Typically, silver copy against a blue background isn't effective. But for this design, it is. Outlining "Plumbing" and airbrushing the bottom half of the letters makes the copy readable and emphasizes the design's focal point. There's no guessing what service this business offers. The oval background ties all the design's objects together. Furthermore, overlapping elements unify the design, so your eyes easily move from left to right.

able distance. The distance from which a sign is viewed determines suitable copy character height. One standard suggests that 1-in. copy can be read from a distance of 50 ft. Only a person with eagle eyes could read something this far away. The rule I was taught states that every inch of character height yields 25 ft. of readability. I believe this rule is truer.

Viewing distance is only one consideration. For on-premise signage, you must also consider the speed at which traffic is moving past the location.

Furthermore, words that comprise initial capital letters followed by lower-case letters are more readable than words that employ all upper-case letters. On the other hand, from a greater distance, signs that incorporate all capital letters are probably more legible.

Outlines and drop shadows can also improve readability. For example, heavy, black, drop shadows can create the right contrast and improve readability. Although black is frequently used for drop shadows, it isn't the only color you should consider. For example, if you're using gold lettering on a bright-red background, use a dark red for the shadow — the dark red will soften the transition from the gold in the foreground to the red background. Popular metallic vinyls aren't always readable.

While contrasting colors can improve readability, contrast in line value, typeface and shape can attract viewers' attention and differentiate your signage. During the paint-and-brush era, script with big block lettering was frequently and effectively employed.

Composition
Good composition — how a sign's elements/objets are organized — improves a sign's visual appeal.

When arranging a sign's design objects, remember a few basic guidelines. Although rules are meant to be broken, I know only a few extremely creative signmakers who can make their own design rules. For the rest of us, it's best to stick to the industry's long-established guidelines.

The first guideline is the KISS rule: Keep It Simple, Signman, or Keep It Simple, Stupid (whichever you prefer). With this in mind, use only three or four key design elements —

pictures; a company's name, logo and/or slogan; trim; and accent stripes and/or borders.

Don't use backgrounds that conflict with the copy, and limit the text to six or fewer words. Typeface variations can be interesting, but don't use too many different fonts. Usually, two different typefaces suffice.

Typically, too many words, colors and design elements create visual clutter. White space should represent 30 to 40% of the overall design. Be careful, though, and don't trap any white space between the design elements — this breaks up the design's continuity and/or flow.

A border frames your layout and ties all the design objects together. In fact, using special-effects films for the border can make your signage even more eye-catching.

After you select the design elements, consider where to position the most important element. If you use special-effects vinyl as a background, make sure it doesn't compete with, or steal attention from, the sign's primary design element.

There are several ways to organize a layout. Some designers divide their composition into quadrants, with the bottom two quadrants occupying slightly more area than the top two. Generally, the design's most dominant element is positioned off-center, but I believe dividing the composition into thirds creates more visual excitement and flair.

Instead of dividing the design space into four parts, use two vertical and two horizontal lines to divide your layout into nine, equally sized rectangles. The primary element doesn't have to be positioned directly in the design's center — a perfectly symmetrical layout can be rather boring and static.

To achieve balance, another important design guideline, you do not have to make your signs perfectly symmetrical. Rather, an asymmetrical layout is usually more dynamic and visually appealing. Arranging a design's key elements on a diagonal axis can convey a sense of movement and energy.

To create more visual interest, position a key element or copy block just above the center. Typically, an item placed slightly off-center attracts viewer attention. Other ways to garner attention include making your primary design element big and positioning an object, such as a pictorial, against or overlapping a sign's edge.

It's also important to consider where viewers' eyes are going to travel after they've seen a sign's primary component. For example, if a design's focal point is a large mug of beer, and you want viewers to notice "Bar & Grill," position the copy next to, or overlapping, the picture. Finally, you can attract attention to an object by setting it apart from other elements.

Most people read a sign the same way they read a printed page — from top to bottom and left to right. With this in mind, when you study a layout, pay attention to a design's key elements. Do your eyes move naturally from one design element to another? Too much space between graphic elements can result in a disjointed design.

A good way to learn about effective composition is to study well-designed ads, signs and billboard layouts, or to visit an art museum and study the works of artists who truly understand composition principles.

In summary, contrasting design elements create more visual excitement. To create contrast, vary the size of words or shapes you use in the design, or use contrasting colors or two different fonts.

If you mess up a design, conduct a post mortem. Where did you go wrong? What could you have done better?

Chapter 24
Screenprinting Basics

Twenty-three years ago, when I worked for a fleet-graphics screenprinter, my training included production art, screenmaking, printing and die cutting. After my indoctrination, I reasonably understood the screen process.

To truly learn screen-printing, I recommend the Screen Print Technical Foundation (SPTF) workshop, "Screen Making: Basic To Professional." At $900 for non-SGIA members, this intensive, one-week course provides a good mix of theory and plenty of hands-on experience.

Screenprinting offers several advantages for numerous vinyl applications, including vehicle markings, window and wall graphics, safety decals and gasoline-pump markings.

In addition, I recommend routinely reading ST's sister magazine, *Screen Printing*, and the *SGIA Journal*, the industry's two best publications.

Educational videos also justify the investment. Lawson Screen Products' website (www. lawsonsp.com) offers numerous screenprinting books and videos. Further web resources include www.screenprinters.net/articles and www.screenweb.com.

The subject of screenprinting often elicits strange looks from signmakers. Sure, screenprinting can be messy and — with solvent-based vinyl inks — stinky. Plus, it involves hazardous materials and expensive and expansive equipment. So why do it?

Although computer-cut and digitally printed vinyl graphics have increased, screenprinting still offers several advantages for numerous applications, including vehicle markings, window and wall graphics, safety decals and gasoline-pump markings.

For high-volume jobs, screenprinting provides significant economies — more durability than digitally printed graphics and less need for overlaminates. Screenprinting also provides closer color matches and richer colors.

For a small signshop, screenprinting can generate high-volume and profitable work, including vehicle graphics, window decals, bumper stickers, POP signage, safety labels, OEM-identification materials and election signage. Screenprinting is typically faster and offers lower material costs and better durability than other vinyl-graphics methods.

Multi-million-dollar companies print most major vinyl-graphics programs, but novice shops can still handle many jobs via simple printing units with minimal financial risks.

The screening process is simple, but printing onto vinyl involves complex chemistry. The vinyl, ink, clearcoat and application tape must be compatible to produce a durable, problem-free product. Strict adherence to manufacturers' recommendations and time-tested practices help minimize mistakes.

The fundamentals

Screenprinting, a form of stencil printing, dates back to ancient times. Now more sophisticated, its basics haven't changed. Ink is still forced through patterned fabric onto a substrate. Ink passes through the patterns' open areas onto the screen. Conversely, the patterns' resistant areas block or prevent ink passage.

The screen fabric is stretched over either wood or metal frames. The frame supports the fabric and serves as an ink reservoir.

A wood or metal frame supports the fabric and serves as an ink reservoir during printing. Signshops prefer wood frames because of their low cost; however, they're prone to swelling and warping.

During printing, ink is poured across the frame's top edge.

A rubber or plastic tool called a squeegee traverses the screen and covers the pattern with ink during the initial flood stroke, or fill pass. A second squeegee pass, called the printing stroke, or impression pass, forces the ink through the screen onto the printing surface.

Originally, the fabric was silk; people called it "silkscreen printing." Today, printers use nylon, polyester and steel fabric.

Patterns can be created several different ways, but fleet-graphics printers generally use the photographic or direct method, in which the screen is coated with photo emulsion. After the emulsion dries, a film positive is placed atop the coated fabric and exposed to a strong, ultraviolet light source. The unexposed emulsion is then washed from the screen, leaving a stencil.

Alternately, the screen stencil can be created using a hand- or plotter-cut stencil film, which is directly adhered to the fabric.

Getting started

The screenprint process usually begins with a sales interview that outlines your customer's needs. For building- and fleet-graphics programs, this typically involves a site survey or vehicle inspection. Take plenty of photographs and make detailed drawings with measurements. If you're quoting an existing program, get decal samples.

The quote request must include the specifics about a decal's finished size, quantity, base substrate, colors and finished cut. Is the decal premasked and clearcoated? Does it require an overlaminate? Determine the project's artwork format, any special packaging requirements and whether sections of the graphic can be run simultaneously or ganged with other pieces.

For a new or modified design, specify the company's corporate colors (PMS numbers, color swatches). What does a client like/dislike about an existing design? What logos, brand names and slogans should be incorporated? What colors, if any, can't be used? For truck and building graphics, what is the design's "live area"?

A typical production order lists the quantity of decals; material required (vinyl product code); sheet size; decals per sheet; finishing requirements; artwork requirements; and whether special color matching, clearcoating, premasking or an overlaminate is necessary.

The production order helps organize shop time and raw-materials orders. Jobs can be completed on time without disrupting other in-house projects. Better production planning should also lead to better purchasing. Combining purchases fosters quantity discounts.

Following job completion, the production order combined with the estimate can evaluate the job. Often, estimates don't reflect production realities.

Film positives and halftones

Screenprinting is characterized by options. Film positives can be hand-cut, computer-cut and produced photographically. Commercial shops typically use film positives.

Traditionally, highly skilled craftsmen hand-cut these films. Subsequently, screenprinters adopted many photographic techniques used in offset lithography. Today, computers generate designs, making screenprinting much easier for signmakers.

The simplest film positive is hand-cut Rubylith® or Amberlith® film. The base layer, a clear film, is covered with a colored film. For hand-cutting, the film is taped over the sample decal, and one film is cut for each color to be printed. Using a sharp knife, the artist uses just enough pressure to cut through the top layer. The base film should only be lightly scored with the blade.

After cutting the film, the artist weeds away the colored excess and leaves a right-reading image. Wherever color film remains, nothing will be printed. Such films are still used today, but they're usually cut by a computer plotter.

A process camera can produce film positives from black-and-white line art, or continuous-tone photographs. These gigantic cameras can enlarge or reduce artwork and photographs. For optimum crispness, oversized artwork is reduced, which minimizes

A form of stencil printing, screenprinting dates back to ancient times. Although now more sophisticated, its basics haven't changed: Ink is forced through patterned fabric onto a substrate; the ink passes through the patterns' open areas onto the screen; the patterns' resistant areas block or prevent ink passage. Whether sophisticated or handmade, all commercial presses comprise a work surface, squeegee and frame.

edge imperfections and improves film resolution. Conversely, enlarged artwork loses edge sharpness.

To print a continuous-tone photograph, an image must be converted into thousands of dots by placing a screen over the negative in the camera. Although both negative and positive films are available for process cameras, negative films are frequently used to shoot the artwork. The negative image reproduces the reverse image, including highlights and shadows.

After processing, the developed negative is placed atop another piece of negative film and exposed in a contact frame to produce a film positive.

Once a screen has been imaged, it can be stored and reclaimed for future jobs.

Before making a screen, carefully examine the film positive with a loop tester — a specially designed magnifying glass. Check for edge sharpness and pinholes. Any pinholes will need to be opaqued.

Because artwork can be computer generated and layout can be output as a film positive, hand-cutting and producing films with a process camera is quickly becoming obsolete.

Fabrics

The screen fabric directly impacts the finished product's quality, especially the print's resolution and ink deposit's thickness. For screenprinting onto vinyl, use a monofilament polyester fabric.

Fabrics with a greater mesh opening (a larger open space between the threads) allow greater ink flow. As with paint, solvents, flow agents and other additives affect ink volume and the ink deposit's dried thickness.

Printing a solvent-based ink onto vinyl generally requires a heavy ink deposit. A coarser mesh (mesh count refers to the number of threads per inch) deposits more ink than a fine mesh. Solvent-based inks typically require a 200-to 280-mesh count. For UV inks, mesh counts of 355 to 390 are more suitable.

High mesh counts create finer detail, but they also restrict ink deposit. Conversely, lower mesh counts allow more ink passage, but less print definition.

Screenprinting fabric colors include white, yellow and orange. Fine-detail work often uses colored fabric. White fabric scatters light; colored fabric absorbs light to prevent it from scattering. Dispersed or scattered light during exposure can destroy fine detail. White fabric, however, suits signage applications because it allows shorter exposure times.

The screen fabric must match the ink system. Polyester fabrics, which are ideal for printing onto vinyl, absorb very little moisture compared to nylon fabrics. This feature also makes them preferable for printing waterbased inks.

A suitable fabric should:

- Be evenly woven for print consistency;
- Feature good abrasion resistance and tensile strength to withstand printing's mechanical stresses;
- Provide chemical resistance to inks, solvents and cleaners;
- Feature fabric construction based on the print substrate, the ink used and required durability;
- Comprise fabric elasticity;
- Feature a low swell rate. Increased humidity swells the fabric and reduces its open area and subsequent ink flow; and
- Include good adhesion characteristics to photo emulsions.

The frame

Whether sophisticated or handmade, all commercial presses have three primary components: the work surface, squeegee and frame. The work surface, or the bed of the press, must be perfectly smooth and flat; work-surface imperfections cause print imperfections.

The wood or metal frame supports the screen mesh during printing. Today, screenprinters use both wood and metal screens. Signshops select wood frames because they're less expensive; however, they're prone to swelling and warping.

Some veteran screenprinters staple fabric to a wooden frame, then stretch it by hand. Although quick and inexpensive, this method's unreliable. You can invest in mechanical or pneumatic stretching equipment. For small shops, I recommend purchasing pre-stretched screens from a distributor. You can purchase a retensionable frame, such as Stretch Devices Inc.'s (Philadelphia) Newman Roller Frame®. I prefer roller frames, because, if the screen tension loosens, you can retension the screen. You can't do this once you've glued or stapled fabric to a frame.

Some primary screenmaking objectives include the following:

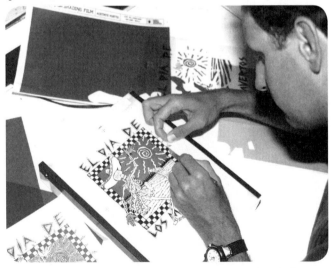

Film positives can be hand-cut, computer-cut or produced photographically. Here, Graphic Designer Jeff Russ hand-cuts Rubylith, one of two simple film positives available.

- Proper fabric tension over the entire screen area;
- Parallel thread direction to each other and with the sides of the frame;
- Enough fabric tension to produce "screen pop," or release, from the printing surface after the squeegee printing stroke; and
- Constant screen tension throughout the entire press run to maintain printing registration.

Improper tension reduces the fabric's mesh openings. Smaller openings will restrict ink flow and the ink deposit's thickness. Improper screen tension can also cause problems with the screen pop or screen snap-off. If the screen doesn't "pop" properly, detail and definition can be lost. Furthermore, improper tension can cause image distortion and misregistration of multiple-color prints.

To compensate for poorly tensioned screens, some printers increase the off-contact distance (the amount of space maintained between the mesh in the frame and the substrate onto which you're printing). Rather than correct the problem, this can actually exacerbate the image's elongation and cause misregistration. As the screen stretches, the stencil in the screen can prematurely break down.

After stretching a screen, check its tension using a tension meter. To ensure accurate measurements and to test for consistent tensioning, take measurements at several places on the screen. Consistent screen tensioning is important to prevent image distortion and misregistration of multiple colors.

Fabric preparation
Fabric preparation — abrading and degreasing the fabric — is critical to creating a high-quality stencil. Improper mesh preparation can cause pinholes and stencil-durability problems.

Abrading the surface improves stencil adhesion, whether it's a direct emulsion or a capillary film. Historically, screenprinters used household cleaners to abrade fabric. However, such cleaners can damage the mesh threads and leave residual particles. Thus, suitable commercial products are better.

Different types of capillary films can be used. One type is cut, reverse-weeded and then adhered to the screen; another type is applied to the fabric, then exposed; and a final type is first exposed, then adhered to the mesh.

Degreasing is the final fabric-preparation step. After the abrasion process, oily residue can still remain in the fabric, and thus decrease adhesion.

Before coating the screen, remove any grease, dust or dirt from the fabric using such products as sodium hypochlorite or bleach. Afterwards, rinse the screen with cold water to remove the residual solution. If cleaned correctly, the fabric surface should have an even film of water.

After degreasing the mesh, dry the screen vertically to prevent dust from settling. Furthermore, coat it after degreasing. Mesh static can also attract dirt.

Screen-stencil creation can involve various products. Coating the screen with a liquid-direct emulsion remains common. Different types of capillary films are also used. One capillary film is cut, reverse-weeded and then adhered to the screen, while another type is applied to the fabric, then exposed. A final type is first exposed, then adhered to the mesh.

Shortly after exposure, thoroughly wash both sides of the screen using cold or lukewarm water with a medium to hard shower spray — never use a pressure sprayer. Afterwards, place the screen in a vertical position in the wash-out booth.

Direct-emulsion method

For the photographic, or direct, method, coat the screen with a light-sensitive emulsion. After the emulsion dries, position the film positive over the coated screen. When the screen is exposed to ultraviolet light, emulsion that isn't covered by the film positive's image hardens. After exposure, the formerly covered emulsion is washed away with water.

Today, emulsions comprise a two-part system: the base and sensitizer. With countless characteristics, emulsions are manufactured to cover various applications.

Unused emulsion sensitizers and bases should be kept in non-metallic containers stored in a refrigerator. Never store such chemicals near heat or bright light, in freezing temperatures, or beyond their maximum shelf life, as they tend to deteriorate and produce poor stencil quality.

Completely dissolve the sensitizer before mixing it with the emulsion base. Strain the sensitizer with a paint strainer to eliminate lumps from the emulsion. For mixing, always use clean instruments, containers and utensils, and mix the emulsion a few hours before using; this disperses air bubbles. Furthermore, remove the emulsion from the refrigerator a few hours before using; allowing it to reach room temperature stabilizes the viscosity.

A sharp image usually requires several screen coatings. Typically, the mesh should receive two coats of emulsion on both screen sides. Data sheets generally specify the number of coatings for each emulsion and mesh count. Although multiple coatings don't dramatically increase stencil thickness, they fill in the recessed areas. Several coatings will ensure a more uniform stencil surface, good print quality and increased stencil life. If the emulsion isn't thick enough, a "sawtoothed" effect — a jagged or stair-stepped edge — can occur.

Screen coating requires subdued light and a dust-free environment. Usually, the last coating is applied to the frame's squeegee side. The coated screen should be completely dried horizontally (the frame's squeegee side facing up) before beginning the exposure process.

Screens dry best in an air-conditioned, humidity-controlled environment. Excessive humidity (more than 60%) inhibits emulsion curing and potentially causes emulsion to break down during printing. Also, because heat and humidity affect coated and stored screens, they should be used within a week.

Dust and dirt on the exposure unit's glass often result in pinholes. Thus, clean the exposure unit's glass before exposing the screen.

Exposure and screen taping

Before exposing the screen, clean the exposure unit's glass; dust and dirt often cause pinholes. To expose the screen, position the film positive on the contact frame. The film's emulsion, or right-reading side, should face upwards so it makes direct contact with the screen's emulsion; then place the frame over the film positive. With the screen's well side, or squeegee side, facing upwards, position the blanket inside the frame — the blanket ensures good pressure and contact between the coated screen and the film positive's emulsion. When the ultraviolet light source is turned on, the exposure process begins.

Shortly after exposure, wash the exposed screen and place it in a vertical position in the wash-out booth. Thoroughly wash both sides of the screen using cold or lukewarm water with a medium to hard shower spray. Never use a pressure sprayer for this process because excessive pressure could destroy the screen's image.

After completely washing out the stencil image, wipe the excess water from the screen using a quality chamois or clean, absorbent newspaper. If the newspaper sticks to the emulsion on the screen's squeegee side, the exposure is probably inadequate. The result may be poor image definition; poor detail resolution; and/or poor stencil, chemical and mechanical resistance.

Another way to check for underexposure is to simply touch the emulsion. If it feels slippery or slimy, the emulsion most likely hasn't hardened properly. During printing, emulsion could deposit in the mesh's open areas and cause clogging. Use an air hose to blow out the open areas and thoroughly dry the screen.

Correct exposure requires an appropriate amount of light and time. Important light considerations include light quality, light consistency and distribution, film-positive quality, emulsion characteristics and various fabrics' detail levels.

After the screen is shot and the image is washed out, apply tape between the frame and mesh to prevent ink leakage from the frame's squeegee side to its substrate side. It's very

The Correct Way to Tape Screens

Using the technique pictured below is guaranteed to save time when taping screens.

1. Cut half way through a 4" to 6" piece of block out tape, creating two flaps. When using split inner tape, remove the liner covering the adhesive on the two flaps.

2. Fold one flap over another to form a corner section.

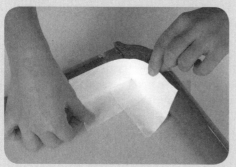

3. Apply the corner section to the corners of the screen. Make sure that the application of the tape is perfectly flat against the mesh of the screen with no wrinkles. Wrinkles will cause ink leaks.

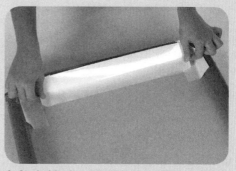

4. Apply block out tape between the corner sections.

5. When using Split Liner™ tapes or Zone Coat™ tapes, tape the top halves of the sections with a small piece of tape.

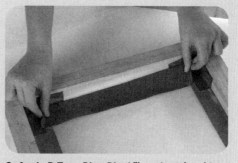

6. Apply R Tape Blue Block™ on the print side of the frame, aligning the tape with the tape applied to the squeegee side of the frame.

important to allow the screen to completely dry. Moisture causes waterbased adhesive to deteriorate, resulting in ink leakage during printing and adhesive residue on the frame and mesh following tape removal.

Use solvent-resistant, block-out tape when masking. The quickest and easiest screen-taping method involves taping the corners, and then the sides. Be sure the tape is completely flat and contains no wrinkles, which encourage leakage. Furthermore, use as few tape pieces as possible so the squeegee won't transfer oozing adhesive pieces to the screen's open areas during printing. This adhesive contamination can inhibit ink flow and cause spotting and pinholes. After applying the tape, coat the screen using liquid blockout between the stencil and block-out tape.

Capillary films

Capillary films — screenprinting stencils that are adhered directly to the screen mesh — comprise two layers: a colored film atop a clear base, which acts as a carrier film. The film helps transfer the stencil to the mesh.

Several types of capillary films are available. Non-photographic stencil films, such as Ulano® Corp.'s (Brooklyn, NY) Amba® film, can be hand or plotter cut and adhered directly to the screen mesh. Photographic capillary films, which replace direct photographic emulsions, fall into two categories: films that are adhered to the screen and then exposed, and films that are exposed and then adhered to the screen.

Whether you're cutting Amberlith® or capillary film, cutting pressure is absolutely critical — the knife blade must cut through the colored layer, but only lightly score the clear carrier. Before cutting the stencil, test the plotter's cutting depth. Cutting pressure is equally critical if you hand cut the film. Today, however, I believe hand-cutting is a lost art.

After cutting the capillary film, weed the areas through which the ink will pass (this differs from cutting Amberlith, in which you weed the areas through which the ink will not pass).

To adhere the capillary film to the mesh, first dehaze and degrease the screen. After the screen is completely dry, inspect the mesh for any dirt or lint. Using a tack cloth, lightly lift the contaminants from the screen. Then place the capillary film onto a clean, smooth work surface, such as a light table, with the film's colored emulsion side facing upwards.

Next, place the screen with the frame's well side facing up on top of the capillary film. The frame should be large enough to provide about 8 in. between any open areas on the film and frame.

On the frame's well side, dampen the mesh area that covers the stencil, using a damp sponge. This wetting procedure softens the film, which causes it to bond to the mesh.

Next, allow the film and screen to dry in front of a shop fan. To accelerate this process, warm the screen and stencil area with a heat gun. Excessively high temperatures can damage the capillary film and mesh; thus, use the lowest heat setting, along with a spreader tip, to diffuse the heat.

After the screen is completely dry, remove the clear carrier film. Before printing, inspect the open areas of both the stencil and film — particles from the capillary film can get caught up in the open areas. Simply remove such particles using a dampened Q-tip.® Imperfections in the film, which allow light to pass through, must also be touched up.

Screenprinting inks

A conventional, vinyl ink comprises three primary components: an insoluble pigment, a soluble polyvinyl-chloride (PVC) resin and solvents. Powdered pigment gives ink its color and opacity/hiding power. The more pigment an ink has, the greater its hiding power. Insoluble pigments are dispersed throughout the ink.

Solid resin acts as the binder, or "glue," in an ink system because it binds the ink to the print substrate. In a conventional vinyl ink, the powdered, soluble resin dissolves in the solvent and affects the ink's gloss level, flexibility and hardness.

A solvent-based, vinyl ink's third primary component is an evaporative solvent, which not only dissolves the soluble resin, but also helps dispense the pigment. The solvent liquefies the ink, and determines the ink's viscosity, flow characteristics and drying rate. Ink pigments and non-reactive resins dry when the solvent evaporates, fusing a colored image to the vinyl film.

To give an ink unique properties, screenprinters can use numerous additives and modifiers. Thinners, retarders, extenders, flow agents and powders accelerate or retard drying, add volume to extend an ink, change the viscosity and flow characteristics, or modify the gloss level. Thinners lower an ink's viscosity. By selecting the right thinner, you can achieve various results.

When choosing an ink, consider substrate compatibility, easy processing, print durability and cost. Also, read the manufacturer's recommendations. Furthermore, keep in mind that in-house testing truly determines the chemical interaction of the substrate, ink, clearcoat and premask.

Durability usually refers to an ink's resistance to water, chemicals, abrasion and fading. Screenprinted truck graphics require ink flexibility; brittle inks can crack and craze when printed vinyl is applied over rivets and corrugations. Processing considerations include the ease of color matching and the ink's drying speed.

Inks are usually classified according to the curing process. Vinyl inks, poster inks and lacquers cure when the solvent, or water vehicle, evaporates. On the other hand, enamel inks cure through oxidation. UV inks cure instantly following exposure to intense UV light, which initiates a chemical reaction. Finally, plastisol inks, which are used in textile printing, only dry when subjected to a short blast of high heat.

In the 1980s, solvent-based inks were frequently used for vinyl graphics. However, UV inks are currently a more common choice. Although UV inks require expensive curing systems, they offer several advantages that outweigh their expense. For starters, UV-ink systems are safer than solvent-based inks that contain volatile organic compounds (VOCs). As the solvent evaporates, VOCs are released into the atmosphere.

UV inks' consistent flow characteristics facilitate printing. Plus, UV inks won't dry in the screens or change in viscosity, making them easier to process. Compared to conveyor belts or drying racks, the UV reactors used in curing these inks take up very little shop space.

Deposits and conditioning

When printing large-format vinyl graphics with solvent-based inks, a thick ink deposit usually produces a more durable print and a richer color. In contrast, heavy UV-ink deposits usually don't cure properly. If UV light can't thoroughly penetrate the ink deposit, the ink will often only surface cure — as a result, uncured ink develops underneath the surface layer.

Several factors govern ink thickness, including screen fabric, the stencil, squeegee type, the printing process and ink system. A coarse screen mesh holds more ink than a finer fabric, and a thicker stencil increases the amount of ink in the screen. Harder, sharper squeegees force more ink from the screen onto the substrate.

Squeegee angle and speed are also important. The squeegee angle shouldn't slant more that 35° from the vertical. Slower squeegee speeds increase the ink flow. Ink viscosity also affects ink deposit. Thinner inks generally flow more readily than thicker inks.

Furthermore, changes in shop humidity affect the moisture content of a vinyl's release liner. By absorbing or losing moisture, the liner grows or shrinks. As the liner moves, the vinyl moves with it. These movements frequently result in color misregistration and/or wavy and curling sheets.

Shops with air conditioning and humidity control encounter fewer problems. Regardless of shop conditions, the best practice is to sheet and condition vinyl prior to production. Conditioning involves racking two sheets of vinyl, face-to-face. Overnight, the liner paper can gain or lose moisture, thereby relaxing and stabilizing the sheet.

Chapter 25
Doming

Lately, our customer-service and sales representatives have been fielding numerous questions about doming. "What is doming?" "What materials can be domed?" "Can you dome printed vinyl?" "What type of doming resin should you buy?" And, finally, "How do you do it?"

This chapter will provide answers to such questions, so you can decide whether doming is suitable for your business, and, if it is, should you attempt doming in-house or subcontract?

One of doming resin's prime attributes is its dimensional appeal for value-added graphics. Here, a Falcon Enterprises (St. Petersburg, FL) employee checks domed vinyl for air bubbles.

Doming is the process of adding a glass-like, plastic-resin bubble to a 2-D surface, which transforms it into an eye-catching, 3-D product. For a small, additional price, you can simulate the appearance of a vacuum-formed or molded part without having to purchase expensive tooling and production equipment.

Vinyl isn't the only material you can dome. Just about anything that isn't porous — metalized polyester, metals, and such plastics as acrylic and polycarbonate — will work. (Note: Anything that's porous can absorb moisture from the air and cause bubbles to form in the doming liquid. I'll discuss bubbles later.)

Recently, one of our United Kingdom-based customers showed me some samples of his work — an award plaque and some prototype signs. In each example, doming resin had been applied over metalized-vinyl films. The resin bubble acts as a lens and magnifies the film's brilliant, light-diffracting effects, creating a very elegant appearance. Even with standard opaque vinyl, doming makes something ordinary appear extraordinary.

Because it's so cool looking, doming has inspired numerous creative signmakers. The most important issue, though, is what doming can do for your business, not what you can do with doming. Doming is simply another way you can differentiate your company from the competition. By converting standard vinyl graphics into rich-looking signs, you can command higher selling prices and add to your profit margins — that's the meaning of value-added.

Resins

Resins typically fall into one of three classes of materials: epoxy, UV-curable resin and polyurethane. Within each classification, several products are available. For outdoor applications, most signmakers and companies that specialize in doming choose polyurethane. After comparing each material's performance characteristics, you'll see why polyurethane is the best choice.

Epoxy resins are popular for doming such advertising applications as key tags, badges and plaques. Available as one- or two-part systems, epoxies are inexpensive and easier to use than other materials. Because epoxies cure slowly, air bubbles trapped in the resin often rise to the surface and disappear, forming a bubble-free finish.

Unlike polyurethane resin, epoxy doesn't react with humidity in the air. Thus, you can work with it without having to run your air conditioner or dehumidifier. In addition, epoxy resins are user friendly and cure at room temperature.

Because sunlight yellows the clear, plastic material, epoxy resins aren't suitable for outdoor use. Plus, due to their softness, epoxy resins can be easily scratched.

Next are UV-curable resins, which fall into two classes: low- and high-intensity. Although low-intensity doming liquids cure when exposed to black light, others require high-intensity UV-curing units.

UV resins work similarly to UV-ink systems. Exposure to UV light initiates a photochemical reaction that quickly hardens the liquid. In contrast, other resin types require hours, and sometimes days, to fully cure.

UV resins are ready-to-use, one-part systems — no measuring, no mixing and no bubbles. Plus, they're relatively safe to use.

Unlike one-part systems, two-part epoxy and polyurethane components must be mixed, which often generates bothersome little bubbles that usually end up in the dispensed doming liquid.

In the future, UV-curable resins might be suitable for high-volume production shops, but not signshops. Here's why: Low-intensity, UV-curable resins harden after being exposed to a black light (the same type of black light common to the '60s and '70s hippie era). Further, low-cure resins don't withstand prolonged outdoor exposure — after six months in the sun, they begin to yellow. As UV exposure continues, these resins cure and harden until they become brittle and crack.

Development Associates Inc.'s (Kingstown, RI) Auto-X casting station applies automotive-grade, polyurethane resin to vinyl. The machine, which operates on an x-axis, comprises a gear-driven, multi-nozzle dispenser.

High-intensity, UV-curable resins cure in approximately 15 seconds after having been exposed to a special light source that costs thousands of dollars. Thus, unless you specialize in high-volume production, these systems aren't for you.

Most signmakers and screen-printers I know use two-part polyurethane systems comprising a resin and an isocyanate curing agent that hardens the doming liquid. Although working with polyurethane resins is typically burdensome, such resins can withstand outdoor rigors. Because they resist yellowing, polyurethane resins are suitable for vinyl graphics and other signage applications, including OEM decorative emblems, nameplates and labels.

In addition, the material remains flexible and doesn't crack, craze or flake. Polyurethane resins not only weather well, but they also block UV light and, in turn, protect the material they cover. This means the base vinyl and printed image will survive longer outdoors.

Earlier this year, I visited the Q Panel testing labs in southern Florida to examine vinyl graphics that had undergone accelerated weathering equivalent to three years of outdoor exposure. The screenprinted and domed graphics panels looked like new.

However, as wonderful as polyurethane resins are, they have one major drawback — they're hygroscopic. They suck up moisture like a sponge. Any moisture absorbed by the doming liquid reacts with the resin and forms tiny bubbles of carbon-dioxide gas. Thus, to minimize bubble formation, you must control your shop's humidity.

Safety tips

Before using doming liquid, read the manufacturer's instructions and the OSHA Material Safety Data Sheet. Because doming liquid can unexpectedly spurt out of the dispensing nozzle, wear safety goggles. If you get some resin in your eyes, immediately flush them out with water and seek medical attention.

To avoid skin irritation, wear rubber surgical gloves; however, never rub your eyes with the gloves. Further, keep food and drinks out of the work area, and don't smoke. Finally, when you're finished working with the material, thoroughly wash your hands with soap and water. Am I starting to sound like your mother yet?

Doming liquid can emit small amounts of slightly toxic vapors, including trace amounts of hydrogen cyanide and mercury vapor. In most cases, this doesn't pose a problem. However, if you breathe enough of these fumes over an extended time period, you could damage your lungs; thus, if you do a lot of doming, ensure your work area is properly ventilated. After all, it's better to be safe than sorry.

Work environment

Controlling a finished product's quality demands controlling your work environment. Companies that specialize in domed graphics, such as Falcon Enterprises (St. Petersburg, FL) or Sunwest Screen Graphics (Winnipeg, Manitoba, Canada), conduct their doming operations in a clean room with controlled temperature and humidity to keep their scrap rates down.

Interviews with production people at both companies produced some helpful tips. If you're interested in doming, consider the following advice.

When you're doming, your work surface and any drying racks must be level, or the resin will run to one side. To ensure its floors were perfectly level, Falcon Enterprises poured new floors before it moved into its new manufacturing plant. I'm not suggesting you pour new floors. However, you should put a level on a worktable and make any necessary adjustments.

Also, keep your work environment air dry and warm. As the air's humidity increases, so do the doming liquid's bubbles. During my first doming attempt, I primarily worried about humid conditions, so I ran my air conditioner for a few days prior to using the doming resin. With dry air, I avoided bubble problems.

To control moisture, some shops use dehumidifiers. Throughout the year, Bill Barnes, VP of production for Falcon Enterprises, maintains a temperature of 70°F and relative humidity of 55%. Other shops keep the humidity as low as 45%.

Temperature, as you might imagine, plays a big part in the curing process. Curing can actually be accelerated at elevated temperatures. Sunwest has a special drying room for this process. My resin manufacturer recommends curing domed parts at 77°F.

Fortunately, I didn't encounter humidity and bubble problems. Rather, dust was a big problem. Before doming, I mistakenly cleaned my floor, which made the dirt and debris airborne. Thus, dust settled on my parts as they were drying. To control airborne dust, floors should be sealed or tiled, walls should be covered with a latex paint, and routine cleaning is essential.

Doming procedures

Several signmakers have mentioned numerous doming problems, including bubbles forming in the resin, parts that won't dry and high scrap rates. But don't let this discourage you because doming isn't that difficult.

A doming starter kit costs between $99 and $250 and includes everything you'll need: a dispensing gun with several mixing needles, a pro-pane flame gun, a repair tool, purge cups and protective gloves. Further, during the doming process you'll need to keep track of time, so keep a clock or a kitchen timer handy.

Starter kits also provide a sheet of adhesive-coated glass. After you weed your graphics, stick the vinyl graphics' release-liner side to the glass sheet — this ensures your graphics remain flat. I needed additional glass sheets, which I covered with paper application tape and spray-mount adhesive. Covering the glass with tape simplifies cleanup. When you're done, just remove and dispose the tape.

For my first project, I domed approximately 50 screenprinted and thermal, die-cut decals the size of a silver dollar. This may not have been a very challenging job, but it was a helpful, first-time doming project.

Numerous vinyl graphics can be domed, including steel-rule, thermal or die-cut decals and plotter-cut graphics. Plotters produce the best results because the edges are clean. Conversely, die cutting can sometimes fracture decals' edges, creating a channel where the doming liquid can spill over the edge.

When you dome a cut-vinyl graphic, you substantially increase the part's overall weight. Consequently, the shear strength of the domed vinyl's adhesive is critical. If it can't support the additional weight, your graphic could slip down the side of a sign substrate or vehicle.

You can also dome various printed graphics, including screenprinted, thermal resin and inkjet. Doming printed graphics requires the ink system to be completely dry. Also, be aware that some inkjet inks are hygroscopic, meaning they readily absorb and retain moisture. Any residual moisture in the ink system can react with the resin and form bubbles. Thus, before charging head-

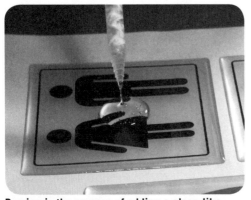

Doming is the process of adding a glass-like, plastic-resin bubble to a 2-D surface to create an eye-catching, 3-D product. This image depicts Chemque's (Rexdale, Ontario, Canada) mercury-free resin, which can be used to build flexible domes for any size sign.

long into production, test a graphic system's components for compatibility.

The dispensing gun in my kit uses cartridges with two chambers — one for the polyurethane resin and one for the hardener. For signmakers, this system is ideal because there's nothing to measure. The two components are mixed together inside the system's mixing needle.

Companies that specialize in doming may choose automatic dispensing machines that can measure, mix and apply precise amounts of liquid onto the part. That's ideal, but such sophisticated equipment costs thousands of dollars, so your volume must justify the investment.

My kit offers a fairly simple process. You just slide the cartridge into the applicator (which is similar to a mini caulking gun), snap off the cartridge's tip, and then snap on the applicator's mixing needle. With the gun tip pointing up, allow the bubbles to rise to the top for a few minutes. As you depress the trigger, the components start to flow and mix together. Do this slowly or the components won't mix properly. Dispense a few milliliters of the doming liquid into a plastic purge cup until no bubbles appear in the resin.

Chemque's Chem-Dec™ doming and coating resins feature a high-build, 3-D doming effect to enhance the appearance of nameplates, decals, automotive badging and trim. Additional applications include signage, electronics potting and molded specialty items.

Be prepared to clean up spills. The first time I used the dispensing gun, the resin shot a couple feet into the air. Fortunately, I was wearing safety goggles. Also, because I had anticipated a mess, I covered my work table with application tape, which enabled me to clean up quickly after completing the project. The process of dispensing the resin is pretty simple — simply pull the dispensing gun's trigger.

Almost any vinyl-graphic shape can be domed. For simple shapes, such as circles, apply a few drops of liquid into the center of the decal, and allow the resin to flow out to the edge and level out. This might require 5 to 10 minutes. A domed part's height can vary from 0.06 to 0.08 in. For more complex or intricate shapes, use the repair tool to coax or guide the liquid to the graphic's edge, or use a metal tool, such as a knitting needle. Generally, it's better to round, rather than square off corners of lettering or graphics. With patience and a steady hand, you can dome some intricate shapes.

Doming liquid's resin has high surface tension, which binds it together. When applied, the resin flows out until it hits the part's edge. At that point, the doming liquid hangs on the edge for dear life. The resin spills over the edge if you apply too much liquid, and the liquid's weight exceeds its ability to hold itself together.

Spillage is rare, approximately 1%, and you can always scrape up any excess with the doming kit's repair tool.

After approximately 10 minutes, inspect the domed graphics for bubbles. Flame-treat the graphic if any bubbles appear. Many available doming kits provide a small, pro-pane-flame gun — the size of an oversized cigarette lighter — with its tip bent at a right angle. The propane torch you may have used for vinyl applications would produce too much heat in this situation.

After igniting the flame gun, hold it 3 to 6 in. from the domed part. A blast of hot air should remove the trapped, carbon-dioxide bubbles. I overdid it the first time — by overheating the resin, I sealed the surface and permanently trapped a zillion little bubbles inside the domed part.

Another trick I've learned but haven't tried is to poke the curing resin with a sewing needle and carefully draw or guide the bubble to the surface. If that doesn't burst your bubble, nothing will.

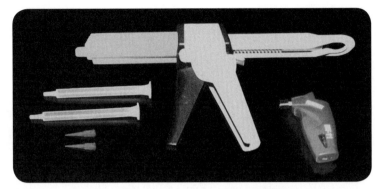

A doming starter kit, which costs between $99 and $250, typically comprises a dispensing gun, several mixing needles, a propane flame gun, a repair tool, purge cups and protective gloves.

Some manufacturers say the resin is dry to the touch in one hour. The best advice, though, is to not touch it for at least 24 hours. If the resin isn't completely dry, you'll likely leave a fingerprint. After the resin is completely cured, it's relatively tough and can even resist mild scratching.

It's usually safe to ship doming liquid to your customers after 24 hours. However, you may want to play it safe and wait 48 hours. Given the type of doming liquid and myriad factors such as curing temperature, humidity and the amount of liquid applied, full curing requires up to a week. Quite simply, more liquid requires longer to cure.

Application is simple for domed, one-part emblems. Simply remove the release liner and stick it in place. If your graphics comprise numerous letters or graphics elements, you must transfer the domed graphics with an application film. Suitable tapes vary. Ask your sign-supply distributor for a recommendation.

Consider the tape's tack level (does the tape have enough grabbing power to transfer the graphics?), whether the adhesive delaminates from the tape's backing and, most importantly, whether the tape leaves an impression on the domed part's surface. With application tape, make sure the part is completely cured before you mask it.

To dome or not to dome

Doming is like any product. Consider: Will the demand for the product in your market justify your investment in tools and resin? How much effort and expense is required to promote the product? What is your likely return on your investment?

To estimate, calculate how much resin you'll need, vinyl costs, how many parts will be lost during production (your scrap rate) and job duration. The $10 resin cartridges are often sold in packs of 10. Resin purchased in bulk can cut costs by as much as 75%. That makes cartridges seem quite expensive, but if you only dome graphics occasionally, the convenience is worth the added cost.

To estimate, you need to determine the size of the part to be domed. A resin cartridge can cover approximately 50 sq. in. You also need to factor in the job's scrap rate. Even if you're good at doming, your scrap rate can be as high as 15%. To play it safe, factor in at least 25%. And finally, you need to add the cost of any remaining material because it may be awhile before another doming customer walks through your doors.

Chapter 26
Digital Printers

Chances are, if you don't already own a digital printer, you're considering purchasing one. Today, an entry-level printing system's price tag — whether it's a time-tested, thermal-transfer printer or a new inkjet system — makes

Flatbed printers like the Durst Rho 160 UV print directly onto rigid sign substrates, thus reducing material costs for vinyl, mounting adhesives and overlaminates. Foamboard is among the many rigid materials such systems can accommodate.

such units affordable for nearly all signshops.

Technological advances, including production-speed improvements, print durability and format size, are additional reasons you should seriously consider going digital. The new generations of eco-solvent printers are especially exciting — they feature exceptional resolutions and good production speeds. Plus, they don't require special, topcoated vinyl.

As exciting as new technology is, before you commit, you should seriously think about which digital-printing system is right for your business. When contemplating such a decision you should consider the following:

Market focus. Focus on which market segments your shop will serve — vehicle graphics, POP, exhibit/tradeshow graphics, backlit signs, banners, window graphics, floor graphics, etc. The choices are endless. Also, it's important to consider digital-printing media types. Cast and calendered vinyl, banner and flexible-signface material, foamboard, expanded PVC, paper and floor-marking film are available. Because you can't satisfy everyone, the key is to focus.

Customer requirements. Depending on the market, customer expectations may vary. Some clients may demand a sharper print than others. Further, a print resolution that's adequate for a vehicle graphic may be unacceptable for a POP display.

Thus, you need to consider your customer's resolution requirements, which involves determining viewing distance. For example, because fleet graphics and outdoor signage will be viewed from distances greater than 50 ft., a 180-dpi resolution may be acceptable. On the other hand, tradeshow posters, which are viewed more closely, require printing resolutions of 760-dpi or higher.

Break-even point. Based on your shop's monthly fixed costs, calculate the sales volume needed to pay for the equipment purchase. To arrive at this break-even point, divide your monthly fixed costs by your average profit margin. The fixed costs associated with buying a digital printer could include the loan payments, equipment depreciation and maintenance contracts, plus a percentage of the rent, utility costs and insurance.

Justifying a printer's cost and all the extras isn't too difficult. For example, if your fixed monthly costs equal $1,000, and you operate your business at a 40% average profit margin, your break-even point is only $2,500.

If your break-even analysis suggests sales won't cover your fixed costs, along with your direct material and labor costs, you have two alternatives: Rework your plan, or outsource the printing until your sales justify an equipment purchase.

Sales forecast. The bigger your investment, the more you need to focus on marketing and sales. In other words, determine to whom you're going to sell before making an equipment purchase. If you don't, you'll probably struggle to find work to fill the machine's capacity. In your sales and marketing plan, include a target and sales forecast that lists how much you'll sell.

To justify a printer's cost, consider your shop's break-even point — simply divide your fixed monthly costs by your average profit margin. The result is the sales volume needed to pay for the equipment purchase.

Start-up costs and cash flow. Do you have enough money to get started and make your monthly payments? How will a printing system's additional expense affect the rest of your business?

If your break-even analysis indicates profitability, crunch more numbers to determine your company's projected net profit and whether you have the available cash to cover your start-up costs and pay your monthly bills.

You should also establish a profit goal. What's an acceptable return on your investment? If your return is less than 15%, the venture probably isn't worth the trouble.

Training, system maintenance and total system cost. Determine the type and amount of system training a distributor provides, the learning curve for your employees and whether a service agreement is available. If the system breaks down, what service does a distributor provide? Also, consider routine maintenance tasks and estimated annual costs for parts replacement and maintenance. When will printheads need to be replaced?

When calculating your fixed monthly expense, consider more than just the printer payment. Other associated costs could include the workstation, software, peripherals, a scanner, digital camera, laminator, ventilation hood, maintenance agreement and training.

Finally, it's important to know a finished graphic's cost per square foot.

Inkjet printers
Inkjet systems print by spraying ink onto a substrate, without making contact with the print. Earlier units offered minimal printheads. However, today's more sophisticated units, with more nozzles, greatly improve production times.

The latest generation of eco-solvent printers boasts outputs as high as 300 sq. ft. per hour. Furthermore, maximum production speeds for the new flatbed printers, with UV-curable inks, reportedly reach 1,000 sq. ft. per hour. Overall, printheads' engineering design changes have improved print quality, and, most importantly, different inks can be used with the new printheads to create greater outdoor durability.

Inkjet printers offer lower capital investment, photorealistic print resolution (180 to 1,440 dpi) and large-format capabilities. Plus, they're compatible with various substrates.

Ink is important because it determines what media types the printer accepts. No universal ink system works with all print media, so it's important to ensure that the printer, ink and media are compatible. Furthermore, ink type determines whether you need topcoated vinyl or uncoated film. If you need to purchase topcoated media, consider what type will work in your printer and how much it will cost.

Printers that use waterbased inks require expensive topcoated media. Dye-based inks' downfall is lack of durability, compared to thermal-transfer and electrostatic printing technologies. Even solvent-based, pigmented inks have limited exterior life, especially if prints are subjected to harsh climates.

Inkjet technology requires drying time, overlaminates and, in many cases, topcoated vinyls. Plus, some systems are limited by slow production speeds. Future inkjet improvements will likely include higher resolutions, increased production speeds and longer outdoor durability.

Continuous flow and DOD

Continuous-flow inkjet systems provide a constant stream of ink droplets. Using positive and negative electrostatic charges, some droplets are deflected into a gutter; remaining droplets are deflected to the print surface. The ink, which is directed into the gutter, is typically recirculated.

Drop-on-demand (DOD) systems, the most common, sign-industry, wide-format print devices, don't use a gutter system to recirculate the ink. Instead, the ink is only expelled from the printhead when needed. DOD systems can be divided into the two following categories: thermal inkjet and piezo inkjet.

Thermal inkjet. This type of printer channels ink into a chamber. The inkjet printhead contains a heat resistor that super-heats the dye-based ink to temperatures as high as 750°F or 400°C. In microseconds, the liquid ink boils and becomes a gas.

In a thermal-inkjet system, the heat resistor turns on and off to control the ink droplets. As the nozzle opens, the heated ink explodes through the printhead onto the substrate. As the waterbased ink dries, it bonds to the surface.

Thermal-inkjet printers typically use dye-based inks, although a few systems can print pigmented inks. Historically, thermal-inkjet systems' intense heat disallowed printing with pigmented inks. The heat structurally altered the pigment, destroying the color, and residue from overheated inks clogged the inkjet nozzles.

Compared to piezo printheads, thermal-inkjet printheads aren't very durable and require routine replacement. Generally, the ink requires filtration to strain out impurities. Without filtration, the printheads clog quickly. Along with ink and substrate, the printheads are considered consumables. (Note: When calculating your fixed expenses, remember to include printhead replacement costs.)

The thermal printheads' consistent orifice size is critical in controlling image output and maintaining quality control. As thermal technology has improved, ink droplets have become smaller, which improves print resolution and provides photorealistic images.

Although inkjet systems are more affordable than other primary technologies, such consumables as printheads and specially topcoated vinyls can significantly add to the finished print's cost.

Moreover, fluctuations in shop environments affect all electronic printing equipment. Although output from inkjet printers remains relatively consistent under temperature and humidity shifts, the ideal signshop environment would include temperature and humidity controls.

For example, Ad Graphics (Pompano Beach, FL) has one room for its inkjet printers and another for its 3M Scotchprint® electrostatic systems. For optimal performance, each room is zoned for its own environmental control.

Piezo inkjet. Piezo-inkjet technology is also popular in the sign industry. In piezo units, inks are mechanically forced through the printhead. The ink doesn't boil. Instead of undergoing four phase changes, it only undergoes two — from liquid to solid. As a result, thicker inks and pigmented inks can be used. Of course, pigmented inks are more outdoor durable than dye-based inks.

Thermal-inkjet systems became popular due to their low purchase cost. And today, many signshops house such units. However, because of the sign industry's demand for more durable prints, interest in piezo systems is steadily growing.

Piezo-inkjet printers work like an electric oil can or squeeze bottle. Ink is drawn from a reservoir into the piezo-electric transducer. As an electric field is applied to the compression chamber, the chamber walls contract and force or squeeze out the ink. Application of another electric charge also causes the chamber to expand, which sucks in new ink to refill the piezo "oil can" chamber.

Among piezo-printing systems, three operation modes are used in printhead design: normal, shear and coupled. In normal mode, the top chamber wall flexes. With systems using shear mode, the two side walls flex. In the combined mode, all three chamber walls flex, which expels the ink from the chamber faster, increasing production speeds.

Although piezo printheads are more costly than thermal-inkjet printheads, they're more durable and reliable, and rarely experience clogging. Piezo systems can use a wide range of inks, including water- and solvent-based, and UV-curable. These systems also accommodate thicker pigmented inks, which yield greater outdoor durability. Furthermore, piezo systems don't subject the ink to super-heated temperatures, and some inks printed using piezo systems dry rapidly, which improves production time.

Flatbed Printers

In the large-format, screenprinting market, and among large signshops, flatbed printers have generated a great deal of interest over the past three years. These systems, which use either UV-curable or solvent-based inks, can print onto a wide range of rigid or flexible substrates.

Because flatbed printers print directly onto rigid sign substrates, material costs for vinyl, mounting adhesives and overlaminates are reduced. By eliminating some production steps, labor costs are also reduced.

Although UV flatbed presses print directly onto uncoated vinyl, special films with permanent and removable adhesives have been developed by such companies as Avery Dennison Corp. (Hamilton, OH). These films comprise a coating that promotes ink adhesion so the printed image doesn't scratch off.

When exposed to intense UV light, ink-system components undergo a chemical reaction that instantly hardens the ink. These systems can print at resolutions of 360 dpi at high-production speeds ranging from 500 to 1,000 sq. ft. per hour. Unlike solvent-based inks,

UV-curable inks don't penetrate vinyl films and attack the adhesive system.

Because UV curing systems can produce ozone and airborne particles, and solvent-based inkjet systems can produce hazardous fumes, both equipment types require a ventilation system. Some printers incorporate a ventilation system.

Systems that incorporate UV-curable inks provide shops with numerous features and benefits. The inks are lightfast, making them ideal for outdoor signage projects.

Because UV inks only cure when they're exposed to intense UV light, the printheads don't clog as readily as inkjet printheads using conventional inks. The systems incorporate piezo-inkjet printheads, which require daily cleaning, and offer a print resolution between 360 and 720 dpi.

The Durst Rho 160, Zund UVjet, NUR Tempo, VUTEK PressVu UV 180 EC, Inca Eagle and 3M Scotchprint® 2500UV are among the UV flatbed printers available.

Not all flatbed printers use UV-curable inks. Solvent-based inkjet systems include the Arizona T220, and Mutoh Toucan Hybrid. Although the resolution of earlier grand-format printers was rather limited, and suitable only for graphics viewed at a distance, today's printers have been greatly improved.

Because some flatbed systems are rather large and hefty, consider the amount of floor space required and whether your existing floor can support the machine's weight.

Eco-solvent printers

Today, eco-solvent printers that can print onto uncoated vinyl films, thus reducing direct-material costs, are making big waves. With pricing starting at approximately $14,000, eco-solvent printers are affordable. Unlike solvent-based inks, eco-solvent systems don't produce fumes. Therefore, no ventilation hood is required. To accelerate drying and promote ink adhesion, the printers use dryers that fuse the pigmented inks to the vinyl media's surface.

The printers' resolutions range from 180 to 1,440 dpi, and they print at speeds up to 200 to 300 sq. ft. per hour.

Although eco-solvent inks are lightfast and adhere well to vinyl, Laura Wilson, Roland DGA Corporation's (Irvine, CA) product manager, recommends using overlaminates, especially for demanding applications. An ink may adhere well to vinyl and withstand UV light, but may not be scratch and chemical resistant. Using overlaminates for such applications as vehicle graphics provides added protection. Some available eco-solvent systems include Roland's VersaCAMM printer/cutter and SOLJET PRO II, and Mutoh's Falcon Outdoor.

Gerber's ELAN™ printers are solvent (not eco-solvent) systems that can print onto uncoated media. The printers' solvent inks are non-toxic and don't contain harmful VOCs. Plus, they're extremely vibrant, and chemical and abrasion resistant.

Thermal-transfer printers

Signmakers interested in adding digital-printing services to their arsenal should seriously consider purchasing a thermal-transfer system. Some examples include the Gerber EDGE®, Gerber MAXX® 2 and Roland's ColorCAMM® PC-600 series printers. These affordable systems are ideal for producing sign applications that require outdoor durability.

Often, thermal-transfer-printer operation is compared to how a typewriter works — both use colored ribbons and transfer color onto a substrate. However, thermal-transfer technology is more sophisticated.

As the name implies, thermal-transfer printers use heat in the printing process. The printheads contain multiple heating elements — each of which rapidly turns on and off to control the printing process. In some systems, the printheads extend the entire width of the print substrate; in other systems, they move back and forth across the web.

Thermal-transfer systems use a cartridge with a printing ribbon. One side of the ribbon is lightly coated with colored wax or resin. During the printing process, the ribbon's coated side makes contact with the print media, and the printhead presses on the ribbon's uncoated side. As the resistors heat up, the coating melts and transfers to the print substrate.

The primary advantages of thermal-transfer systems are brilliant colors and extended outdoor durability (UV stability and water resistance). Many spot colors are available, which makes matching corporate colors easy and accurate.

Thermal printers are also easy to operate and maintain. According to John Mazzeo, owner of Classic Signs in South Plainfield, NJ, "Thermal-transfer printers are low maintenance — they require daily cleaning, which involves wiping down the printheads with isopropyl alcohol and cleaning the sprockets."

Although the cost of printing cartridges may seem high, thermal-transfer printing systems don't require expensive topcoated print media. Also, spot colors allow for the use of a single color, reducing print costs.

Overlaminates may not be required for many jobs, but they're recommended for demanding applications. The finished product's competitive cost and good outdoor durability make thermal-transfer systems viable digital-printing options for signmakers.

Thermal-transfer printers also produce relatively photorealistic output. For example, the Gerber MAXX™ 2 can print at resolutions up to 300 dpi. Production speeds for this printer are reportedly as high as 110 sq. ft. per hour, when printing one color. When printing four-color process work, production speeds are approximately 24 sq. ft. per hour.

The EDGE 2 prints nearly 300 sq. ft. per hour for a single spot color and nearly 60 sq. ft. per hour for CMYK. Coupled with minimal drying times and other post-print processes, thermal-transfer throughput is tough to match.

Electrostatic printers

Contrary to many industry predictions, electrostatic printing still exists — and for good reason. "When solvent-based inkjet systems came on the market, I thought their introduction marked the beginning of the end for electrostatic printers," said Rich Thompson, president of Ad Graphics (Pompano Beach, FL).

He continued, "I fell in love with inkjet technology and immediately purchased a printer for my shop. However, after working with the solvent-based inks, I realized the technology's drawbacks. Today, to print fleet graphics, vehicle wraps or outdoor signage, we use our Scotchprint® 2000 electrostatic printer."

According to Thompson, the advantages of electrostatic printing include high-production speeds, excellent outdoor durability and reliability. He also discovered that the heavy solvents used in his shop's inkjet printer attacked the vinyl's adhesive. The solvent inks also required the installation of a ventilation hood. After subjecting prints to outdoor weathering tests, Thompson believed Scotchprint graphics outperformed inkjet prints.

Electrostatic printers operate similarly to a Xerox® office copier. Printing is usually a two-step process. The image is first printed, in reverse, on a special paper. Then the image is transferred from the paper, onto a substrate, using a lamination process. The new generation of electrostatic printers streamlined the process by directly printing onto pressure-sensitive vinyl.

Electrostatic technology works on the idea that opposite, electrical charges attract. The system's printhead comprises thousands of electrical wires, which deliver, or deposit, electrical charges to the paper. As the paper passes over pigmented particles with an opposite electrical charge, the particles are transferred to the paper. The paper is then laminated to the substrate, and, during a heating process, the particles are deposited and fused to the substrate.

It doesn't matter that the electrostatic prints' resolution (200 to 400 dpi) isn't as high as inkjet prints, because large-format electrostatic prints for fleet graphics, banners, outdoor advertising and construction-site signage are generally viewed from a distance.

Electrostatic printing's major disadvantage is system cost, which can range from $100,000 to $250,000. Electrostatic systems also require constant air conditioning and humidity control, because electrostatic toner is sensitive to changes in humidity. Because electrostatic graphics are unappealingly dull when printed, they require lamination or clearcoating.

Inkjet media

In an ideal world, every inkjet vinyl would work with every inkjet printer, and purchasing decisions would be based on price comparisons. However, this isn't the case. Selecting inkjet vinyl is difficult because photorealistic reproduction, color gamut and outdoor durability rely on the interrelationships of printer, ink, vinyl and (when used) overlaminate.

Media selection becomes even more confusing with the increasing number of media manufacturers entering the sign market. Although several printer manufacturers and media producers have formed business partnerships, and developed consumables warranty packages, whether a particular inkjet print will survive outdoors is often unknown.

Because the technology is constantly changing, many real-time test results are simply not available yet. The only way to find out if a pressure-sensitive vinyl, ink and overlaminate are compatible and durable outdoors is to subject the combination to outdoor exposure in the intended area of use.

Vinyl and inks are usually tested in demanding environments located between the latitudes of 30° N and 30° S. Test sites in coastal areas, at higher elevations or in areas of significant chemical pollution subject a graphics system to the most demanding environmental conditions. Thus, such outdoor exposure is usually the most reliable test.

The secret is the topcoat

Inkjet vinyls differ from standard films due to their topcoating systems. These water-soluble topcoats absorb and bind the ink to the substrate. Topcoats also speed up ink-drying times. Without this special surface treatment, tiny ink droplets bleed together, destroying any print definition. With untreated vinyl, the inks would also wash off with the first sprinkling of rain, or quickly abrade with a light fingernail scratch. Each topcoating is specially formulated to match the substrate and work with an ink's chemistry.

The most complex topcoating systems comprise two layers. In a dual-layer topcoat, ink is passed through a top layer and absorbed into a base coat. The top layer acts as a barrier, protecting the ink system from UV light, water and abrasion. The coating must also be absorbent enough to accept heavy saturation.

In a perfect marriage between ink and film, ink droplets are drawn directly down into the topcoat, where they're absorbed and prevented from spreading. Good drop integrity results in crisper images and more intense, vibrant colors.

In contrast, when the chemistry between the ink and coating is wrong, prints can be plagued by dot gain — a spreading or bleeding of the ink on the surface of the topcoating, resulting in washed-out colors and fuzzy images.

In evaluating products, remember that all topcoated vinyls aren't created equal. The system's chemistry, the thickness of the coating and the consistency of the finished film all affect print quality, durability and print-production waste.

Also important are overlaminates and clearcoats. Regardless of which printer you select for your shop, the printed image can often be easily damaged by abrasion and chemicals. Some overlaminates even help block UV light, slowing the fading to which inkjet prints are susceptible.

Testing an overlaminate is important to ensure compatibility with the other components and assess a print's overall durability. Unlikely as it may seem, an overlaminate's adhesive could react adversely with the ink system.

For your consideration, when evaluating inkjet vinyls, buyers should:
- Compare the color richness from one print to another.
- Look for fuzzy edges or loss of detail.
- Determine the time it takes for the print to become dry to the touch.
- Consider the vinyl's adhesive system and whether it's aggressive.
- Consider repositionable and removable adhesives.
- Find out about any effects the liner paper may have on the printing process.

In comparing print quality, such statistics as resolution and dot size only tell part of the story. When testing different inks on the same vinyl, the results can vary greatly. A graphic printed with pigmented inks can look significantly different when printed with dye-based inks.

Pigmented inks have improved UV resistance. However, their color gamut is limited. Dye-based inks are bolder and brighter, and you can achieve more realistic pictures with visual punch.

Some topcoatings merely bind the ink to the substrate, while others enhance the ink's appearance. Two different coatings on the same vinyl will produce dramatically different results.

Sign-supply distributors, and printer and vinyl manufacturer representatives, can provide guidance in selecting the right product for the job. This is what value-added sales is all about.

Conclusion

To justify your investment in a new printing system, ask yourself if the equipment will help generate enough additional business to provide you with an acceptable return on your investment.

You should purchase a new printer if it will help you increase sales among existing customers and/or pursue new markets. Ensuring that your investment pays off requires proactive marketing and selling. Today's new digital-printing equipment may be the sign industry's better mousetrap. However, the world will never beat a path to the signmaker's door just for the sake of the latest and greatest technology.

Chapter 27
Ecosolvent Printing

Ecosolvent printers have been a real godsend for the sign industry. The printers' low price tags are a major reason why they're so popular. Not suprisingly, an estimated 60% of signshops anticipate purchasing ecosolvent printers.

Printer setup is rather straightforward. Approximately two hours after unpacking the equipment, you're ready to print. Plus, there are no dangerous fumes or odors, and no ventilation system is required. And, if you follow a few basic rules, you should have few, if any, problems.

If problems occur, they typically involve poor ink adhesion, shrinkage, vinyl curling, color shift and/or dot gain. For trouble-free results, use the right material, operate the equipment in a suitable environment and use the proper print settings. This month's column offers suggestions for printing onto vinyl using an ecosolvent printer.

The right stuff

The latest generation of ecosolvent inks allows you to print onto various uncoated, monomeric and polymeric calendered vinyls and cast-vinyl films. However, this doesn't mean you'll get great results with every vinyl.

You can't print onto every type of vinyl. Actually, you can only print onto a limited range of uncoated vinyl films. With some vinyls, ink adhesion is a problem. For this reason, prior to production you must test and evaluate your vinyl choice with your printer and inks.

To choose the right material, consult the printer manufacturer and distributor. Because many printer companies have performed extensive vinyl tests, they can tell you which films work best with their systems. Plus, they have profiles for the recommended films. As a rule of thumb, vinyl films with a matte finish print better than high-gloss films. Furthermore, films with polyethylene-coated liners are generally less prone to curling than films with clay-coated liners.

Prior to printing, wipe down the vinyl using a towel moistened with Rapidtac or isopropyl alcohol to remove dust and other contaminants, such as oils and plasticizer.

A dry, clean and dust-free surface is thus ready for printing

Control your shop environment

Although ecosolvent printers are designed to function in an office or signshop, you'll achieve the best printing results if you operate in a clean, temperature- and humidity-controlled environment.

Ink adhesion largely depends on ink and film chemistry, but other variables exist. Two different printers using the same equipment, ink and vinyl film can produce different results. Humidity, temperature and cleanliness also affect printing results.

For consistent results, you must control your shop's humidity and temperature. Manufacturers recommend their optimal humidity and temperature conditions. When printing onto vinyl, shop temperatures should be between 65 and 85°F (18 and 30°C); relative humidity should be between 30 and 70%.

The lack of environmental control can contribute to color shift, which explains why some shops can print a file on different days and achieve different results. Companies that specialize in digital printing often invest in an environmental-control system, which maintains consistent temperature and humidity year round.

Shop conditions can vary from one season to another, especially when you change from air conditioning to a heating cycle. If your shop's temperature is too low, and/or the humidity is too high, the ink won't wet out the surface and sufficiently bind the vinyl for good ink adhesion. Instead, the ink will pool on the surface and form little islands. This condition occurs more frequently with heavier ink deposits. Extremely cold or hot shop environments can contribute to ink-adhesion problems. Humidity control is also important. If relative humidity is too high, you could experience dot gain. If the humidity is too low, it could contribute to clogged nozzles.

Machine settings

Ecosolvent printers can produce high-quality prints for such durable signage applications as fleet graphics. Their resolutions range from 180 to 1,440 dpi.

Because machine settings for printing onto vinyl will vary, you must establish a profile for each film. The profile defines the ink, material, printer speed and resolution. The best profile will yield the optimal print quality based on a given set of variables. Remember, if you change one of these variables, you must change the profile. Printing variables account for color shift as much as fading is caused by the sun's bleaching effects.

Any setting recommendations should be regarded as guidelines, not hard and fast rules. When printing onto any vinyl, "test, don't guess" before you go into production. When establishing a profile, print a test swatch with ink lay-down from 0 to 100%, generally at increments of 10%.

The newest generations of ecosolvent inks are designed to work with the printers' multiple heating phases. In the preheating phase, the vinyl is heated to prepare the film's surface to accept the ink. With ecosolvent printers, you can adjust the printer temperatures from 40 to 122°F (35 to 50°C). Preheating opens up the micropores on the vinyl's surface, which gives the film more "tooth" for the ink to bite into. The heating also helps prevent dot gain and prohibits the ink droplet from contracting to maintain dot integrity.

To achieve acceptable adhesion, the ink must flow into the pores. During printing, the ink is heated, which helps it wet out on the film's surface. After printing, the curing phase fixes, or locks, the ink into place to bond the ink to the vinyl. For the ink to cure, the solvent must evaporate. The process is time and temperature dependent. In colder shop temperatures, the ink doesn't flow as well onto the vinyl's surface, and often, the ink doesn't cure properly.

Inadequate ink curing can affect some vinyl films' adhesion. Remember, the ecosolvent ink comprises a solvent. If the solvent doesn't evaporate from the ink, it can penetrate the vinyl and attack the film's adhesive layer. High heat can cause the ink to dry too fast and clog the nozzles.

Ink primarily comprises the carrier or vehicle, which is typically either water or solvent. As the ink cures, the carrier evaporates — it shrinks. By removing the large volume of solvent, mechanical tension is created. In this curing process, the ink contracts, which contributes to the vinyl's contraction.

The heat used in ecosolvent printers contributes to the vinyl's contraction or shrinkage. Heat affects any material — this is especially true of vinyl. During the printing process, the vinyl is subjected to several temperature changes. On the roll, the vinyl is at room temperature; then the film is pre-heated. Under the printheads, the film is subjected to even higher heat. After printing, the vinyl is heated again to cure the ink. Finally, the vinyl cools back down.

As the vinyl is heated in the printer, it expands. As the vinyl cools down, it contracts. If the film expands and contracts at a much greater rate than the liner, tunneling could occur. Because elevated temperatures can contribute to shrinkage and curling, some signmakers prefer to lower the heat settings.

Heavy ink concentration, which occurs when you print a solid color or drop shadows, also contributes to shrinking and curling. Drop shadows create great contrast between lettering and a sign's background and produce a 3-D look. But when the shadow bleeds to the edge of a cutline, the vinyl can curl. Thus, to minimize curling, try to provide a 1/4-in. or 6mm outline around the printed image.

Heavy ink concentrations can cause other problems; they can take forever to dry. The ink can also bleed or puddle into ink islands. A slower printing speed also reduces the likelihood of vinyl contraction because the ink has extra time to start drying before another ink layer is printed. The downside is less printing efficiency.

During heating, the expanding vinyl can tunnel on the release liner. Tunnels between the vinyl and release liner can form as the heated film under the dryer expands and pushes against a cooler, rigid, film mass.

When printing onto vinyl, use lower-resolution settings. On glossier films, this will help prevent the dots of higher-resolution prints from running together. When printing higher resolutions, adjust the density curve in your profile to control the amount of ink that you print. Also, slow down your print speed. You'll experience less bleeding, especially when printing onto glossier vinyls. You also give the ink more time to dry before the printheads make another pass. Generally, you get better results with matte-finish vinyls as opposed to glossy ones.

Vinyl graphics don't usually need a high resolution. With such outdoor applications as fleet graphics, prints are viewed from a distance. In such cases, lower-resolution graphics provide more contrast and, consequently, are more readable.

Finishing the job

Printer manufacturers claim that overlaminates or clearcoats aren't necessary for most signage applications. At the very least, digital prints should be lightly sprayed with an aerosol clearcoat. Before weeding, spray the print with a uniform coating, then weed the print. You can apply application tape as soon as the clearcoat is dry — usually within 15 minutes.

You can use either a water- or solvent-based clearcoat. Spraying a medium coating of clear is recommended, but don't overdo it. Heavy coatings may cause the ink to run. However, you really have to overdo it to damage a print. Waterbased clearcoats usually dry faster than solvent-based ones. To my knowledge, a waterbased, clear, aerosol version isn't available. You can apply clearcoat with a brush; however, in most cases, the coating is heavier than it should be.

Some solutions for preventing curling include printing at lower resolutions, lowering the heat settings, laminating your application tape soon after printing and cutting, and applying the graphics as soon as possible.

Chapter 28
Lamination Tips & Tricks

There's usually more than one way to do something. For example, signmakers have devised many different techniques for laminating prints. If something works, continue to do it; if it doesn't work, try something new.

In this chapter, I'll share some tips that have worked for some industry friends. Hopefully, these tips will help you prevent laminating mishaps, save production time and, most importantly, improve profitability.

Safety and compatibility

If you take safety seriously, review the safety precautions outlined in the equipment manual with your employees. One of the most important rules is to never wear loose clothing when operating the machinery. If you wear a tie, tuck it in your shirt. Better

Because all laminators are different, you must determine your machine's optimum settings. Once you've loaded the overlaminate roll onto the unwind shaft and secured it in place, ensure that the roll can't shift from side to side. The tension on the unwind roll should be low — basically, use the least amount of tension to accomplish the job.

Laminating prints is usually easier if one person stands in front of the machine and feeds the print, and another person guides the laminated print to prevent the laminate from wrapping around the rollers. Apply laminating film to a print in one continuous pass, and never stop the process in the middle of a print to check its progress. Doing so will pick up the releaser liner's impression, and cause silvery lines in the adhesive and over the print.

yet, lose the tie. I once forgot this rule and nearly laminated a tie. Fortunately, I was spared an embarrassing accident. Although safety gadgets and features are usually built into laminating machines, they don't always activate when they should.

Furthermore, long hair should be tied back. Wearing a hair net could prevent hair from falling onto a print and being encapsulated. Another important rule is to keep your hands away from any moving parts. Fifteen years ago, my right hand's pinky finger was crushed in a machine. Although the injured finger has healed, it doesn't look the way it did before the accident.

If you're wiping down the rubber rollers, disconnect the power — this prevents accidents from occurring, especially if the laminator is accidentally turned on. If you need to advance the rollers, reconnect the power. After you've advanced the rollers, pull the plug again. Use a clean, lint-free rag and a mild, non-abrasive cleaner to wipe down the laminator.

The interaction of inks, printing systems and print media involves very complex chemistry. Incompatibility among components can adversely affect the overlaminate's adherence to the print and cause delamination. The residual solvent in the overlaminate's adhesive

could react with the ink system and cause a color shift. Always test and evaluate your raw materials (overlaminate, inks, print media and mounting substrate) before a production run. And, when you find a winning combination, stick with it.

I recommended using a calendered-vinyl overlaminate with a calendered-vinyl film and a cast-vinyl overlaminate with a cast-vinyl film. Similar films expand and contract at the same rate. If they don't, the

Some people argue that with a little heat on the top roller — from 80° to 110° F — the overlaminate's adhesive flows out better, thus creating a better bond to the print. Conversely, others argue that heat can cause the laminating film to shrink and delaminate from the print. Overall, it's best not to use heat, unless you have a serious problem.

overlaminate can delaminate from the base film or substrate. A tunnel could also form between the two films.

Overlaminate problems can also result when a print is mounted to a substrate that expands and contracts at a high rate. The resulting tension between the print and laminating film can cause tunneling or delamination.

Storage and environment

Overlaminates and other pressure-sensitive materials, such as application tape and vinyl, need to be properly stored. Don't stack rolls of unboxed overlaminates horizontally on top of each other. If the rolls aren't suspended within the box, they may get flat spots, which can appear as a visible line after the overlaminate is applied to the print.

Better yet, store rolls in their boxes, stacked upright, until you need them. Although the overlaminates' corrugated containers are designed to withstand shipping's rigors, horizontally stacked boxes can collapse under excessive weight.

The best storage area is the same one in which you print. The ideal temperature and humidity control is 65 to 72°F at 40 to 50% relative humidity. Under these conditions, overlaminates' expected shelf life is more than a year.

In addition to air conditioning and humidity control, your shop needs to be spotless. When cleaning your machinery, never use an air hose, because the dirt becomes airborne and eventually settles on your raw materials. Dirt trapped between the print and overlaminate usually tents the laminating film over the trapped particle. To clean your laminator's rollers, use a tack cloth.

Drying and threading

Before applying an overlaminating film to your print, be sure the ink is completely dry. Wait 24 hours, so the inks can cure properly. High humidity can prolong the drying period. Even if the print feels dry to the touch, the ink might not be completely cured.

Print-drying time varies, depending on the print's ink density, the ambient temperature and your shop's humidity. Usually, inks dry more slowly when temperatures are cooler and humidity is high. Consult the manufacturer's recommended drying time. Furthermore, before trimming, mounting or rolling up the laminated print, lay it flat for 24 hours.

Before threading the overlaminate through the machine, study the roll-laminator diagram in your owner's manual. The laminated print's finished appearance depends upon several variables, including the feed roller's unwind tension, the nip rollers' pressure, laminating speed and temperature.

All laminators are different, so you must determine your machine's optimum settings. After you load the overlaminate roll onto the unwind shaft and secure it in place, make sure the roll can't shift from side to side. The tension on the unwind roll should be low. As a general rule, use the least amount of tension to accomplish the job.

Don't allow the unwind roller to free-wheel. A slight amount of tension ensures that the overlaminate feeds into the nip rollers evenly. Some back tension will prevent wrinkling and trapped air bubbles.

High unwind tension stretches the overlaminate. Any stretched film, especially a stiff film like polyester, has a memory and tends to return back to its original shape. This can cause the laminating film to either curl in the direction of, or delaminate from, the print. Excessive tension can also contribute to delamination of the print from the substrate to which it's mounted.

High unwind tension and high laminating pressure facilitate print curling. The signmaker who gave me my first lamination instruction said that, when his prints are curled, he tells customers the graphics have been "pre-rolled" for shipping.

Lamination
As I previously mentioned, signmakers employ various laminating techniques. Kapco Graphic Products provided the following helpful technique: Slit the release liner, 18 to 24 in. from the overlaminate's lead edge, across the web, without cutting through the film facestock. Don't remove the liner from these first couple of feet of overlaminate. This lead edge will aid film feeding and prevent the adhesive from sticking to, and wrapping around, the bottom roller.

Next, from the slit you've made, peel back enough liner so you can tape it to the take-up shaft (if your laminator is equipped with one). As you guide the overlaminate through the rollers, ensure that the material is feeding evenly, and that no wrinkles or bubbles are forming. During startup, you likely wasted a few feet of overlaminate.

Once everything is running smoothly, you can feed the prints into the laminator. Laminating prints is always easier if one person stands in front of the machine and feeds the print, and another person guides the laminated print and prevents the laminate from wrapping around the rollers.

Apply laminating film to a print in one continuous pass. And, whatever you do, don't stop in the middle of a print to see how it's going. Starting and stopping during this process will pick up the impression of the release liner — at the point where the liner and film separate — causing silvery lines in the adhesive and over the print. This is especially noticeable over the print's dark "shadow" areas. Note that, in most cases, this problem can be easily fixed by burnishing the line with your thumb nail to aid the adhesive in wetting out.

Speed and settings

Pressure-sensitive films usually laminate better at slower speeds. Thus, set your laminator's machine speed to 3 to 5 ft. per minute. Then set the machine's pressure-control gauge, which governs the nip rollers' pressure.

Pressure-sensitive overlaminates require pressure so the film's adhesive properly flows out and makes complete contact with the print's surface. Insufficient pressure can result in silvering and air-bubble formation. Conversely, excessive pressure can cause print wrinkling and curling.

Requisite pressure for applying a laminating film varies. Overlaminates generally require between 30 and 50 psi of pressure. If air bubbles appear, you can increase the nip pressure, slow the laminating speed, and, if worse comes to worse, apply a little heat.

Thicker overlaminates, such as 5-, 10- and 15-mil "mar-resist" vinyl or polycarbonate, typically require higher pressure settings. When laminating graphics with these films, using a "sled" (which comprises an 1/8-in. sheet of polycarbonate or acrylic) underneath the print can increase pressure and even out the pressure across the laminator's web.

Should you use heat when laminating a pressure-sensitive film? Some people argue that with a little heat on the top roller — from 80 to 110°F — the overlaminate's adhesive flows out better, which creates a better bond to the print and can prevent silvering.

An opposing argument says that heat can cause the laminating film to shrink and delaminate from the print. Heat can cause other problems too, including waviness, tunneling and wrinkles. Thus, it's best not to use heat, unless you have a serious problem.

Rolling and cleaning

Whenever possible, store and ship prints flat, especially when using polyester overlaminates. Generally, a flexible overlaminate, such as vinyl, is preferred for rolled graphics. If you need to roll a print, place the printed image on the outside of the roll. Never roll a print too tightly, or you'll end up with tunneling between the overlaminate and print.

When rolling up a printed vinyl film with a vinyl overlaminate, the core's outside diameter should be no less than 6 in. When rolling graphics, which are protected with a thick (5 to 15 mils) vinyl or polycarbonate overlaminate, the interior roll should be no less than 12 in. in diameter. Basically, the thicker the overlaminate, the larger the core diameter.

Before delivering prints to customers, instruct them how to properly care for their new graphics. To clean laminated prints, use a mild, non-abrasive cleaner — one that doesn't contain a lot of acid or alkaline, or a strong solvent.

When washing graphics, use a sponge or soft rag. Hard, bristle brushes will scratch an overlaminating film's — especially 3-mil PVC films — surface. To dry graphics, use a soft, clean towel.

Section 4
Vinyl Application
& Removal

Chapter 29
Safety First

Photo courtesy of Arlon

One careless moment can cause a life-changing accident. A friend of mine — one of the most experienced decal installers in the industry — recently cut his wrist severely. The 4-in. gash required hospitalization and resulted in a $35,000 medical bill. In addition to medical expenses, my friend was out of work for eight weeks. Fortunately, he's gradually regaining the use of his hand. I'm glad to say that when I saw him last, he had a firm grip on a bottle of beer.

While there are more dangerous occupations, the vinyl-graphics industry has its share of hazards, including illness resulting from long-term chem-

Safety should be a paramount concern of all vinyl shops. Even though the manufacturer's guidelines didn't call for gloves, Jim plays it safe and wears gloves when working with a chemical vinyl remover. Jim also uses gloves when working with a pressure washer.

ical exposure, repetitive-motion injuries and accidents caused by the unsafe use of equipment. Most accidents are preventable if we make safety a workplace priority.

Chemical management

When you work with chemicals, it's important to understand the hazards involved. Also, be familiar with first-aid procedures in the event of an accident.

Chemicals can enter your body several ways. They can be absorbed through you skin and eyes, or their fumes can be inhaled. Also, solvent-contaminated food can be ingested.

Careless handling of certain chemicals used in substrate preparation or vinyl and adhesive removal can cause a wide range of health problems, including respiratory ailments, liver and kidney damage, corrosive burns and damage to the central nervous system. Also, remember that solvents are usually flammable.

Some decal installers wipe down trailer surfaces with xylene-saturated rags. This chemical not only enters your system through inhalation, but it can also be absorbed through your skin. When working with xylene, consider the possible long-term effects to your central nervous system, kidneys and liver.

Some adhesive removers contain solvents, such as toluene, which require special handling. Toluene, the chemical in airplane glue, can cause intoxication. Many years ago a friend and I removed graphics using an adhesive remover that contained toluene. We handled the job during wintertime in Canada in a facility with inadequate ventilation.

The adhesive was saturated with the chemical remover, which transformed it into a jelly-like substance. Each trailer had more than 128 sq. ft. of adhesive that required removal.

Within 15 minutes, we were higher than a kite. Imagine two men over 6-ft. tall and more than 240 lbs. laughing and yelling loudly while jumping up and down on flimsy scaffolding consisting of two ladders and a 2 × 12-in. plank. It wasn't a pretty sight, and we were very lucky the plank didn't break.

Some chemical removers can cause skin irritation, allergic reactions and, in some cases, corrosive burns.

About 20 years ago, I tested a vinyl decal remover designed for use on graphics applied to unpainted metal. This was one of those "too good to be true" products. The remover came in an aerosol can, and, after I sprayed the chemical remover, the vinyl and adhesive washed off the surface with nothing more than a garden hose. I don't remember if the chemical was acid or caustic, but I'll never forget how dangerous this chemical could have been if not handled properly.

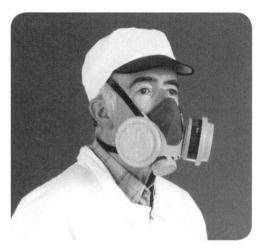

When working with chemicals that emit noxious fumes, gas masks are a must.

Before testing the remover, I carefully read all of the manufacturer's recommendations — wear safety gloves, protective aprons, rubber boots, air respirator and safety goggles. I decided that I would pick and choose which safety recommendations to follow. I would learn the hard way not to compromise safety precautions.

I decided to test the aerosol chemical outdoors for proper ventilation. What I didn't anticipate was that the wind would suddenly shift directions, blowing the atomized mist back into my face. I was very lucky that the only consequence was a tingling sensation as the chemical burned my skin. The scariest part was that some of the remover burned a 2-in.-diameter hole in my flannel shirt. It caused a similar burn on my torso, eating away an outer layer of skin after turning it a mahogany-brown.

Words to the wise
Based on my experiences, here are some common-sense recommendations for safely handling chemicals:
- Maintain an up-to-date file of all of pertinent information provided by the manufacturer, such as the Material Safety Data Sheets (MSDS), product bulletins and application instructions. Make sure you read and understand these sheets, then explain to your employees the potential hazards of handling a particular chemical, as well as how to safely use and store these materials.
- Invest in appropriate safety equipment, such as chemical gloves and aprons, air respirators and rubber boots. Safety equipment may not be stylish, but it beats brain damage, chemical burns, chronic illness, birth defects and other potential dangers.
- Lead by example. You can't expect your employees to follow safety guidelines if you don't act responsibly yourself.
- An effective safety plan can't be a part-time activity. Rather, make it part of your daily routine. At this year's International Letterheads Meet, a signmaker from California told me his company has instituted weekly safety meetings and requires thorough annual physicals of its employees. Two great ideas!
- Keep a record of any accidents and review the results at your safety meetings.

Don't play with fire

Installers frequently use propane torches and torpedo furnaces, which can pose serious fire hazards. I'm proud to report that, to date, I haven't burned any trailers down. Others, however, haven't been so lucky. Unsafe practices can cause trailer fires, so be sure your insurance plan covers accidental fires.

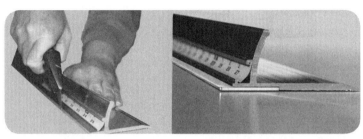

When cutting vinyl, using a safety ruler can prevent a lot of accidents and cut fingers.

The risk of fire increases when flammable chemicals are used near an open flame, such as one from a torch. When you're not using a bucket of cleaning solvent, put a lid on it to reduce the risk of fire and explosion. Even though fires are rare, always have a fire extinguisher handy.

Repetitive-motion problems

Some repetitive motions used in vinyl graphics installation — such as squeegeeing or burnishing vinyl with a rivet brush — subject muscles, nerves and tendons to unusual stresses. Some of the more common injuries include tendonitis and carpal-tunnel syndrome.

I know one female installer who developed carpal-tunnel syndrome after weeks of squeegeeing. This injury can be painful, resulting in weakness of the hand muscles. To immobilize her wrist and minimize inflammation, she was forced to wear a splint for several weeks.

Until I experienced tendonitis myself, I believed repetitive-motion injuries were a scam contrived by lawyers. Last January, when I participated in an application test of vinyl on a riveted trailer surface, I tried to keep up with two guys about half my age. As I was working, I felt a sharp pain in my elbow. Rather than admit that my body can't do what it could 25 years ago, I continued to work. Wisdom doesn't always come with age; I'm still coping with recurring tendonitis.

Many injuries are avoidable if you exercise common sense. Vary your work activities, take frequent breaks and, if an activity causes pain, stop.

Ladder and scaffold safety

In the mid-1970s, I learned my first lesson in ladder safety while working as a construction manager. We were constructing a large office complex next to another building.

I soon learned that OSHA occupied the adjacent building. One of the agency's inspectors welcomed us to the neighborhood by citing us for numerous violations. Although I was bitter about the citations for years, I later realized that some of these safety regulations made sense. They include the following:

- The top of a ladder must extend at least 3 ft. past the last step. This provides space to safely step on or off the ladder.

- Inspect ladders, planks and scaffolding for damage before each use. Defective equipment should be discarded and replaced immediately. I know of one case in which a 12-in. plank snapped in half from the weight of two hefty installers. Luckily, no one was hurt.
- Invest in sturdy equipment. After you set up ladders or scaffolding, securely position the legs of the equipment on level ground. Lock the scaffolding wheels in place. Once, while I was installing graphics on a truck-canopy fascia, my scaffold toppled when one of the legs rolled off the pavement. Thankfully, I jumped clear of the fallen scaffold.
- Wear non-slip shoes, and keep the job site in order. You can easily slip and fall on a slick release liner that's been carelessly discarded on the ground.
- This may sound silly, but don't allow any horseplay. I know of a case where one installer, playing around, caused another to lose his balance and fall, resulting in a broken arm.

Taking the time to institute a safety plan can be rewarding for your company in many ways. Fewer accidents and a healthier workplace will pay off in improved employee morale, less time lost from work and lower insurance premiums.

Chapter 30
Tools of the Trade

With the right pressure-sensitive film and tools, you can decorate just about anything — riveted and corrugated trucks, concrete or textured walls. While professional tools won't automatically make you a professional, you can't do a professional job without them.

Rivet brushes are invaluable tools for burnishing vinyl films onto rivet heads, as well as applying vinyl onto corrugated or textured surfaces.

Many tools are relatively inexpensive and readily available from a distributor. Some distributors even offer vinyl-installation tools in a pre-packaged kit. The best such kit is Jay Lansburg's Installation Kit for Sign Artists.

My personal vinyl kit includes rivet brushes, plastic squeegees, felt squeegees, an air-release tool, low-friction sleeves and an assortment of cutting tools. This compact kit has everything I need and fits in my suitcase.

Some subtle differences between tool brands exist, so try various products and use what works for you.

Squeegees 101
Used to burnish pressure-sensitive vinyl, a squeegee is little more than a thin, flat, 3 × 4-in. piece of plastic. Nevertheless, it's an installer's most important tool. Although squeegees look alike, their hardnesses can differ.

Soft squeegees nick and wear out easily, and often generate tiny bubbles in the vinyl. Thus, professional installers generally prefer hard, nylon squeegees. They cost more ($2.50 to $3.50 each), but they last longer than their flimsy counterparts, and their stiffness forces air from the film.

When applying vinyl graphics to corrugated or contoured surfaces, however, a softer squeegee usually works much better. With the right pressure, the more flexible squeegee conforms better to the curves.

Before starting, inspect your squeegee's edges, which should be smooth and straight. Using squeegees with nicked, bowed or uneven edges often causes bubbles. Most squeegees with nicks or burrs can be sharpened by vigorously rubbing the edge of one against the head of another. Sharpening should become a habit.

European installers often prefer felt squeegees, which are sold in different densities, shapes and thicknesses. At a price range of $8 to $10, the denser and harder felt squeegees are usually preferred for vinyl applications. My advice is to try numerous squeegees and use what works for you.

Low-friction sleeves
After applying vinyl and removing the application tape, resqueegee the entire graphic, especially the edges to prevent adhesion failure and edge lifting. Without the protection

of application tape, hard squeegees can easily scratch bare vinyl.

To protect vinyl, use a low-friction sleeve. Comprising reinforced paper, these sleeves slip over a squeegee and cost only 20¢ to 25¢ each. Another way to prevent scratching is to wrap a squeegee with heavy cloth or application tape.

Brayers and roller applicators

I've never been a fan of brayers or roller applicators because I work faster and generate more pressure with a squeegee. Still, many signmakers swear by them.

Roller applicators allow the installer to apply firm, downward pressure when installing pressure-sensitive vinyl to smooth, flat surface. I prefer the Sable block roller because it's easy to handle, and I think it delivers the most pressure.

Rivet brushes

This stiff-bristled, nylon brush burnishes vinyl films onto the heads of truck rivets. However, it's also a handy tool for applying vinyl onto corrugations and textured surfaces such as banners with a heavy scrim, concrete, stucco or cinderblock. The most common rivet brush has a 1-in.-diameter head, although 3-in. brushes are also available. One-in. rivet brushes cost $6 to $7, while the 3-in. brushes, which have shorter, stiffer bristles, cost $15 to $20.

3M™ recently introduced a rivet brush with a thicker, more comfortable grip. In my opinion, working with a thicker-handled brush puts less stress on the tendons around your elbow to reduce the likelihood of tendonitis. I also like 3M's brush because its stiffer bristles provide the most pressure.

Installers prefer a stiffer brush, so they frequently trim the bristles to a shorter length. And some installers carry several brushes with bristles of varying lengths. The shorter, harder-bristle brushes are used for warm-weather applications, when vinyl is softer, more pliable and less likely to crack under pressure. As temperatures become colder, vinyl becomes harder, more brittle and apt to shatter. Under these circumstances, brushes with longer bristles should be used.

A utility knife is one way to cut bristles to a desired length. Hold your knife steady against the bristles and rotate the brush with your other hand. After this procedure,

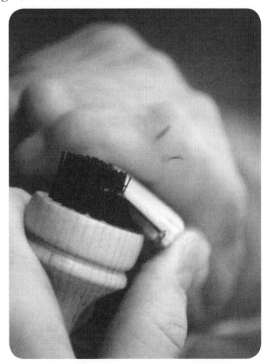

Frequently trimming rivet brushes provides stiffer bristles for the installer — and requires less elbow grease. Wear safety goggles to protect your eyes from flying bristles.

the bristles will be sharp enough to damage the vinyl. Quickly pass a propane torch over the sharp points after trimming. The flame's heat will round the bristles' tips.

Save your old squeegees and rivet brushes for removal work. Squeegees coupled with a chemical adhesive remover can be used to scrape softened adhesive from a truck surface or sign substrate.

Rivet and film cutters

For certain applications, film manufacturers recommend cutting vinyl around the base of each rivet head. Cutting vinyl around rivet heads on stainless-steel trailers — an absolute must — is an example. Reflective sheeting used for conspicuity striping must also be cut. Uncut vinyl often lifts around rivets, becoming brittle and eventually cracking.

Rivet-cutting tools make this job easier. Snapped onto the tip of a soldering gun, the heated tip cuts through the reflective sheeting with a slight twisting motion. These tools usually cost approximately $11 each. Film-cutting tools are available for cutting vinyl around 13⁄32-in. and 1⁄2-in. rivets and cost approximately $37 each.

Torches and heat guns

When applying cast vinyl over rivets, corrugations or textured surfaces, film must be heated to 500-700° F. An industrial heat gun does the trick. (A hair dryer, however, does not.)

Most professional installers, however, prefer a propane torch. Its heat is more intense — allowing faster work — and you don't need electricity. Just be careful not to burn the vinyl or ignite nearby flammable or explosive material.

I prefer a torch to a heat gun, but both tools should have a place in your toolbox. Your toolbox should also include extension cords and replacement propane bottles. Self-igniting propane torches, which are convenient to use, are sold in a kit for approximately $25 to $30. A good industrial heat gun will cost at least $85.

For removal work, a weed burner saves time. This torch can produce a 2-ft. flame and 100,000-150,000 BTUs to quietly soften the vinyl and adhesive.

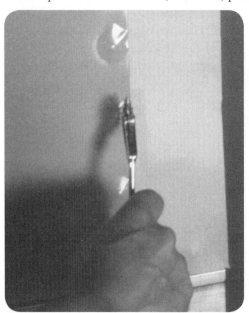

A Wartenberg pinwheel — a tool chiropractors use to test nerve response — creates tiny punctures prior to burnishing rivet heads, allowing air to escape.

Knives and scissors

Any good toolbox should include an assortment of high-quality knives and scissors. A sharp knife is essential for cutting vinyl at the seams. Uncut vinyl tears when the panels expand and contracts as temperatures change.

Olfa makes a popular variety of knives with snap-off blades, available for $8 to $12. Invest in a good pair of scissors, which will cost $15 to $25. For cutting vinyl, I prefer the Olfa OLO Rolling Scissors, which use roller bearings to cut. The rolling scissors cost approximately $20, and produce a clean, straight cut.

Straight pins and air-release tools

Air bubbles trapped under vinyl are common occurrences. The small ones dissipate with time. However, a pin should be used to puncture large pockets. Force out the air with your thumb — never use a razor blade or Xacto™ knife to puncture the bubbles. A razor or knife blade creates a slash in the vinyl, which will open in time. On the other hand, the hole created by a pin prick will close around itself.

Some air-release tools feature a retractable pinpoint. Replacement pinpoints are available from sign-supply distributors. Prices range from $5 to $45.

Wartenberg pinwheel

Vinyl that covers rivet heads must be punctured with several pin pricks prior to burnishing with a rivet brush. Many professional installers use a Wartenberg pinwheel, a chiropractor's tool used to test nerve response.

By rolling the needle-sharp spikes on either side of the rivet row, the vinyl is sufficiently punctured to allow air to escape as the material is burnished around the rivet head. Medical-supply stores — as well as some sign-supply distributors — sell this tool for approximately $25.

Layout tools

For calculating dimensions on an installation diagram, an architect's scale rule, proportion wheel and pocket calculator are essential. A tape measure, straight edge, T-square, felt-tip marking pens, Stabilo pencils and a chalk line help position the graphic elements. (Don't apply pressure-sensitive film over chalk lines because the dust will contaminate the adhesive.)

To tape graphics in place, have plenty of 1-, 2- and 3-in. masking tape. Wider rolls of masking tape are also necessary if you use hinge-application methods. Twenty-five-ft. CenterPoint tape measures, which cost approximately $20, are also helpful when laying out a job because the lower scale on the tape indicates the center.

Edge sealer

Some graphic applications are prone to edge lifting. For example, frequent spillage from gasoline tankers subjects their decals to extreme abuse. Painting a thin line of clear edge-seal prevents lifting and peeling.

Commercial edge-sealers usually come in 1-pint cans. A 1/4-in. paintbrush gives good control in applying a smooth coating of sealer without drips and runs.

Remember, edge-sealer should only be painted on the edge. Never paint vinyl graphics with varnish. Varnishes contain very hot solvents that can penetrate vinyl facestock and attack the adhesive system.

Pint cans of a commercial edge-sealer typically cost approximately $15. Two-part edge-sealers are recommended for reflective sheeting. If a commercial edge-sealer is not available, you can substitute a screenprinting clearcoat.

Application fluid
I've done very few wet applications because they're usually unnecessary. Appying vinyl to acrylic, however, requires application fluid. I recommend using a commercial application fluid such as Rapid Tac®, Action Tac or Splash. Manufacturers of these fluids produce a consistent product.

Adhesive removers
When I performed removals, I used an assortment of chemicals, including kerosene, lacquer thinner, xylene and citrus-based solvent cleaners, such as Orange Peel Adhesive Remover or Rapid Remover.

Be prepared
While working for one screenprinter, I scheduled and organized applications. After experiencing many job delays because I inadvertently left some tools behind at the shop, I compiled a tool checklist. Double-checking this list became part of my routine before hitting the road. The result was that jobs started on schedule with less down time.

One last tip: Don't forget to bring the graphics.

Chapter 31
Conducting Job Surveys

Imagine that a prospect with a fleet of 20 van trailers walks through the door, ready to hire you on the spot. He gives you all his specs — sizes, colors, pictures and even the competitor's pricing.

There's a temptation to take the order, no questions asked, and sweat the details later. After all, when a client is ready to place an order, isn't it your job is to make it easy for him to buy? Why not take the order?

Read on for some answers.

Vehicle surveys reward you with sales if you have the creativity to provide solutions.

Plan ahead

As tempting as it may be to pick the low-hanging fruit, shortcuts in the sales process can result in costly problems.

An onsite vehicle survey is critical because it affects many aspects of the job — design, material selection, estimating, production planning and installation. Important details can be overlooked when you rely solely on customer-supplied information.

Job surveys are especially valuable when planning and scheduling graphic applications for programs where facilities and vehicles are in multiple locations. For fleets like these, your scheduling requires that you determine the number of vehicles and when they are available.

Thorough examinations enable you to spot conditions that could cause vinyl failures. Then you can engineer solutions. Cement-truck graphics, for example, are subjected to regular splattering of caustic cement, along with washing solutions containing a mild acid. Without the protection of an overlaminate, their printed markings are quickly damaged. Markings for such vehicles should be edge-sealed.

Eyes and ears open

A survey's success depends on the quality of both your questions and the answers. Compile a checklist of questions prior to the survey. For a proper survey, have a camera and tape measure, as well as a notepad and pencil or tape recorder.

To meet a company's design objectives, ask targeted marketing-related questions which get to the heart of these goals. A company's president or marketing director is usually a good source for this information.

Budget is critical: How much money has the prospect budgeted for the program? You should spend less time with a contractor with 10 vans and a $2,000 budget than with a bread company with 20 van trailers and a budget of $40,000.

Here are some suggested marketing questions:
• What are their sales goals?

- What advertising and marketing programs will be implemented to achieve those goals?
- How do customers, employees and (if applicable) stockholders view this business?
- How would they like to be viewed by the public?
- Who are their competitors?

Color is critical to any program; obtain color samples from the company if possible. If vinyl swatches are not available, get the Pantone numbers.

Determine whether liberties can be taken with the existing color scheme. To approve color matching, the customer should examine the color sample under lighting similar to how it will normally be viewed. Colors appear different in sunlight than under fluorescent lighting.

Similarly, before attempting to reproduce an image as either an enlarged screenprinted decal or digital print, evaluate the customer's expectations and explain the technology's limitations. Supply the designer with corporate logos, typefaces and slogans, along with reprints and information concerning advertising and marketing programs.

For new designs, photograph the vehicles or building and provide the designer with the necessary sketches and dimensions of the useable space.

Approximately 15 years ago I was installing graphics for a pizza chain's "show" trailers. The graphics were designed to cover 75% of the trailer with reflective sheeting. After taping the markings to the vehicle's side, we realized that they wouldn't fit. Luckily, we modified the design to make it work. A salesman's negligence caused what could have been a disaster because he guessed the trailer's dimensions rather than measuring.

To ensure that markings fit a vehicle, many fleet-graphics companies produce full-size paper drawings or production art before manufacturing. By taping a paper drawing against the side of a truck, you give the customer the opportunity to see the scale of the graphics and ask for changes.

Artwork approval provides you with a signed record in the event of questions and disputes. Make sure that the person who approves designs and colors has the ultimate authority to sign off on the job.

Look for land mines

Pay careful attention to the vehicle's obstructions, such as locking bars, mirrors, windows, louvered vents and rub rails. Hopefully, the designer will avoid these obstructions.

During your inspection, note vehicle details, such as surface smoothness, rivets, corrugations or exterior posts.

Check for rust, peeling or chalking paint, and surface damage requiring repair. By recommending corrections, you will distinguish yourself as a professional rather than just another shop that prints vinyl graphics.

If the substrate will be painted just prior to installation, determine the paint's curing time. Curing times vary and are affected by the environment. A polyurethane paint can take three days to completely cure under normal conditions. During winter, however, that same system could require weeks to completely outgas. Tip: To check whether or not a paint system is fully cured, use the thumbnail test. Press your thumbnail against the painted surface. If your nail leaves an impression in the paint, it isn't fully cured.

Questions regarding the removal of old vinyl graphics and the application of the replacement vinyl graphics are important. In most cases, the customer will want you to provide these services, but in some cases, the customer has his personnel perform the work. You need to know whether they will require training.

Repeated chemical spills commonly attack tanker graphics. Vinyl installations for these vehicles require edge sealing, while overlaminates are essential for printed graphics.

Environmental concerns

Before choosing materials or manufacturing methods, examine the environmental conditions to which the graphics will be exposed. If the vehicles are kept in service for only a few years, a less-expensive calendered vinyl may be an appropriate choice.

If you're developing a graphics program for a tanker fleet, find out if the graphics will be exposed to spillage. Acids and caustics will rapidly erode a screenprinting clearcoat and then attack the ink system.

Even pigmented vinyl is not immune to chemical deterioration. Chemicals can leech the plasticizer from the PVC facestock, making it brittle and prone to cracking. Tanker graphics usually need an overlaminate and require more time because the graphics should be edge-sealed.

Improper cleaning causes some graphic failures. Some cleaning solutions for fleet vehicles contain harsh chemicals, such as hydrofluoric acid.

Sometimes the fleet operator's cleaning system causes problems. Municipal buses are frequently washed daily with gigantic nylon brushes that subject graphics to a beating. In a few months, these brushes can grind a clearcoated print down to the base vinyl. Bus graphics washed in this manner should be overlaminated.

When bidding on an existing program, learn how the existing graphics were fabricated.

If the graphics were screenprinted, ask for a set of the markings. You may be able to "nest" the graphics and save on materials usage. If vinyl samples are not available, tracings can eliminate guesswork.

As an alternative to tracing paper, try using a premium-grade paper with an ultra-low-tack adhesive. Squeegee the mask, and, with a soft drawing stick, rub over the edges to reproduce the image.

When surveying, note whether the graphics are pigmented vinyl, screenprinted, painted or digitally printed. Also determine whether they have a clearcoat or an overlaminate and whether they are computer- or die-cut. Indicate the type of vinyl used: cast, calendered, polyester, metallic or reflective.

If job specs have been written, obtain a copy, along with any design boards and installation prints. Take plenty of photographs to document graphic mistakes such as fading or shifting colors, as well as vinyl failures such as edge peeling, tenting around rivets and channeling in corrugation low points.

Hopefully the time that you invest will reward you with greater success in selling fleet graphics.

Chapter 32
Surface Preparation

One reason that vehicle graphics fail is poor substrate preparation. While some general rules apply to surface cleaning, be aware of the exceptions to avoid potential pitfalls.

Basic procedures

Don't be fooled by how clean a truck or car looks. A vehicle surface attracts dirt like a magnet. This is the reason vinyl manufacturers recommend applying graphics immediately after surface cleaning.

Using a non-abrasive cleaner the day before graphics applications is the best first step for preparing a truck surface for vinyl.

Inspect vehicles prior to installation. Check surfaces for peeling and chalked paint, as well as rust spots. These conditions must be corrected before the surface can be considered vinyl-ready. Typically decal installers contractually require fleet owners to complete the first washing and will charge extra if it's not done.

Cleaning a vehicle surface is a three-step process. The day before vinyl application, vehicles should be washed using a non-abrasive commercial cleaner. This allows the unit to dry completely, even underneath rivet heads and between panels.

A vehicle must be completely dry before graphics can be applied. Otherwise, moisture under the panel seams will likely cause lifting around the edges. Water trapped underneath rivet heads can cause the vinyl to form "tents" in these areas. Over time, tenting vinyl becomes brittle, then cracks and breaks away from the surface. Moisture trapped between the film and substrate can prevent adhesives from bonding or freeze in the winter and cause adhesion failure.

Before installing vinyl graphics, pass a propane torch over rivet heads and panel seams. Condensation will bubble from under the rivets and expel any dirt trapped under the rivets or panel seams.

After washing, completely remove grease, soot and tar on the unit. I suggest using a petroleum-based cleaner, such as Du Pont's 3919S Prepsol®, available at most automotive-supply stores. Stronger solvents soften and dull painted surfaces, especially repainted jobs, so test solvents on an inconspicuous area of the vehicle before usage. This is also a good practice when using solvents to remove adhesive from old graphics.

If you use petroleum-based cleaners, which leave an oily residue, wipe the surface with isopropyl alcohol. Saturate a clean rag with solvent and scrub the surface. Before the solvent dries, use clean paper towels to wipe away the dirt. To slow the evaporation process on hot days, some installers mix alcohol with water.

Working with new paint

What could be better than a newly painted factory finish? That may have been the case in days gone by, but some trailer manufacturers now use slick clearcoats, similar to those that cause vinyl-adhesion problems with automotive paints. Some powder-coated paint formulations can also be difficult for graphic applications.

Learn as much as you can about the paint system when working with repainted vehicles. Each paint formulation is different — some contain such additives as wax and silicone, both of which can cause adhesion problems.

Allow a paint system to dry for one week before installing

A clean surface enhances the quality and lifespan of vehicle graphics such as this one, an RV promoting a California radio station. Supergraphics (Sunnyvale, CA) created this wrap.

graphics, especially with polyurethane paints. Curing time can vary depending on the temperature, humidity and amount of hardener used. In cold conditions, curing can require weeks.

Applying graphics over a paint system that's not fully cured can result in paint outgassing, which causes bubbles to form underneath the vinyl. Thus, time is the only remedy for outgassing.

My advice is to avoid guesswork — apply a piece of polyester film to the surface and wait a few days. If bubbles don't form, you can proceed.

Out with the old

As paint weathers in sunlight (UV light attacks and degrades the resin that binds the paint ingredients), it oxidizes and gradually erodes, leaving a powdery, white residue.

Chalking should not cause alarm — it's the result of the normal aging process. All oil-based paints chalk. Excessive chalking, however, can be a problem, often caused by excessively thinned paint. Very light coats of paint and poor quality paints are also more prone to excessive chalking.

Typically, chalking is not an issue because powder is flushed from the vehicle surface during regular washings. However, if chalking remains on the substrate, it must be cleaned prior to graphic applications to prevent a vinyl failure. Chalk contaminates the adhesive and forms a barrier between pressure-sensitive adhesive and the paint, preventing a good bond.

The degree of chalking and the substrate's condition differ between vehicles — strict rules for substrate preparation don't exist. That's why chalking removal may require experimentation.

Some experts suggest using a power washer to flush the surface with clean, warm water, while others contend that even high-pressure power washing only removes a minimal amount of residue.

If you opt to use a power washer, exercise extreme caution. Excessive water pressure can literally blow paint off a trailer panel. Be sure to keep the tip of the washing wand at least one foot from the surface.

Don't try to wash chalk away with a solvent cleaner, especially a petroleum-based product such as Prep Sol™. A chalky surface is a layer of dead paint, which will absorb the solvent like a sponge. Rather than cleaning the surface, the solvent will turn the white powder into a white paste, which will merely smear around the paint surface. Some chalk and dirt will be driven back into the paint, and the chalk will float back to the surface, causing adhesive failure.

Abrading the surface is often the best — although certainly not the easiest — solution for cleaning a chalky substrate. Many decal installers use a kitchen scouring pad, such as a Scotch-Brite™ pad, and water to clean the surface. This will require plenty of clean rags to mop up the considerable mess.

Scouring pads effectively abrade chalk from the surface, but they will also scratch. In most cases, the result should be acceptable, considering that excessive chalking isn't aesthetically pleasing in the first place.

Some installers use polishing or rubbing compounds to remove chalking. Be sure to select one without wax because such cleaners leave a residue that can compromise the adhesion.

I suggest powdered cleaners with mild abrasives. In addition to removing the dead paint, these cleaners are said to scuff the paint, providing more surface area to bond the adhesive.

Surfaces 101
Here are some suggestions for treating various vehicle-graphic surfaces:

- Unpainted aluminum oxidizes rapidly. Older aluminum truck surfaces that are badly pitted or oxidized must be degreased and etched with a commercial acid brightener, such as Alumaprep®.

   Oxidation can also be removed by scrubbing the surface with a Scotch-Brite™ pad and an abrasive cleaner, such as Comet®. After this process, clean the unpainted aluminum with isopropyl alcohol before applying the vinyl.

- Stainless steel. In my opinion, nothing looks classier than a stainless-steel trailer. Its polished, silvery surface looks great and weathers well. However, stainless steel is a notoriously difficult surface for vinyl applications. One screenprinter I worked for turned down all jobs involving this surface.

   The main problem is that vinyl won't adhere to rivet heads. Some say that, because a cheaper grade of stainless steel is used in making rivets, grease collects in the metal's microscopic pores.

   Installers have tried everything to remedy the problem — including trying to burn the grease with a propane torch — all to no avail. This is why film manufacturers recommend cutting the vinyl at the base of each rivet head. Rivet-cutting tools make this job easier.

- Fiberglass. Fiberglass body components, such as hoods and fenders, provide a smooth surface for graphics application. However, problems can occur. In casting these parts, the mold is coated with a release agent before the gel coat is applied. Occasionally, the release agent transfers from the mold to the finished surface.

    Inadequate cleaning of this coating can contaminate the adhesive of a vinyl graphic. To remove this waxy coating, prep the surface with a reducer (I suggest DuPont's 3812S) followed by a final cleaning with isopropyl alcohol and water.

    Keep in mind that new fiberglass can outgas. Fiberglass is fully cured by the time graphics are applied, but low temperatures and high humidity lengthen the curing period. Resin that's not fully catalyzed also slows curing and outgassing.

    Fiber-reinforced plastic (FRP) panels are often used for the sides of straight trucks. Comprising an outer layer of gel coat and an inner layer of resin impregnated with a woven fabric, these panels are laminated to a 3/4-in. plywood core. FRP panels are prone to outgassing and stress cracks in the gel coat. Cracking often occurs near the top and bottom rails. Wherever the gel coat cracks, the vinyl film does as well.

    One company's solution was to screenprint or apply graphics to .040- and .063-in. aluminum, then screw the designs into the FRP panels. FRP panels can chalk, and they should be treated the same way as chalked paint.

Plastic-clad doors

Most roll-up trailer doors are clad with painted metal, which is a great surface for vinyl graphics. However, in years past, some doors were clad with such plastics as polyethylene and polypropylene. These plastics have very low surface energy and don't allow an adhesive to flow or spread out properly. When this happens, the vinyl's adhesive doesn't form a good bond with the substrate.

To create a bond, flame treating can alter the surface energy of polyethylene or polypropylene by oxidizing the molecular structure of the surface, thus permitting adhesion.

First, clean the surface using isopropyl alcohol. Using a propane torch with a spreader head, wave the flame over the spreader head's surface. The blue flame, which is the hottest part, should touch plastic. Don't worry about heating the substrate. If the flame stays in motion, this treatment shouldn't even warm the surface.

To test whether the surface is properly treated, sprinkle a small amount of water on it. If the water forms droplets, the substrate requires additional treatment; if the water forms a smooth film, the surface is ready to accept vinyl.

Over and under

In most cases, old vinyl graphics should be removed before installing new graphics. Removing vinyl is usually a miserable, time-consuming job; in some cases, it makes sense to overlay old graphics rather than remove and reapply.

If you go this route, remember two simple rules. First, only overlay an old graphic if it's in good condition. Applying vinyl to a cracking or peeling decal guarantees failure. Second, make sure the new graphic completely covers the old one by at least 1/2 in.

A word of caution: Don't expect vinyl film manufacturers to warranty such applications.

Chapter 33
Application Over Rivets

If you've ever observed a professional installer applying large-format vinyl graphics, you've probably noticed that he works quickly and with seemingly great ease.

The photos accompanying this article demonstrate the installation of a large, 3 × 12-ft. decal. Remarkably, the installer completed his work in less than 20 minutes — without the aid of application fluid — on a hot, humid, summer day, and the finished product was flawless.

Be sure to puncture vinyl properly, using a tool such as a Wartenberg pinwheel or mutli-pin air releasing tool. Don't use a knife, which will create slashes that will eventually open.

Why does a seemingly straightforward job frustrate many signmakers? Because, to a large extent, successful installers use professional grade tools, work with user-friendly materials, have received training and possess years of experience.

The right tools for the job

The type of vinyl and application tape that you select for a fleet-graphics job can determine an installation's difficulty. Vinyl with an aggressive, pressure-sensitive adhesive can be a nightmare to apply. In summer heat, this adhesive often adheres too soon, sticking with little or no applied pressure. And once it sticks, you're stuck. Repositioning the graphic at this point generally results in distorted or damaged vinyl.

In contrast, films with repositionable adhesives, such as 3M™'s Controltac™, Avery Graphics' EZ Apply™ or Arlon's ProFleet™, are much more forgiving. They allow the applicator to move or reposition vinyl on the surface until firm pressure is applied. Low initial tack and encapsulated adhesives are two types of repositionable systems.

Another relatively new, easy-to-handle adhesive system is 3M's Comply™, which features a network of tiny air channels embossed on the adhesive's surface. These channels allow air to escape from beneath the film. This feature enables even novices to perform wrinkle- and bubble-free vinyl installations to challenging substrates, such as corrugated surfaces.

When choosing a vinyl, consider that adhesives' systems' recommended temperature ranges vary. Some films that apply easily in hot temperatures may not have enough tack to stick on a cold winter day, or vice versa. The manufacturer's spec sheet or bulletin indicates the temperature range for a particular film series. Ideal temperatures range between 60-80°F. By selecting the right film for the job and using a few time-tested tricks, you can apply vinyl in conditions from 40-95°F.

Although manufacturers have made significant improvements with calendered films, most professional installers still prefer cast vinyl, which is naturally soft and conforms to rivets, textures and corrugations. Because very little mechanical stress is generated in the manufacturing process, cast film is dimensionally stable and less prone to shrinkage. The stable plasticizer used in its formulation keeps it soft and pliable for years. By comparison, some calendered film becomes brittle and cracks with age.

Using heavyweight, premium-grade application tape makes installations easier. Approximately 30% thicker than standard-grade application tapes, it adds "body" to flimsy films and facilitates handling large sheets of pressure-sensitive graphics. Premium tapes can be a lifesaver during outdoor installations. When the wind kicks up, heavier transfer tapes are less likely to tear if the graphic flaps furiously in the wind. That heavier body results in smoother vinyl with fewer wrinkles during squeegeeing. After completing an application, premium tapes can be removed in one sheet, not in bits and pieces.

Vinyl education

Years of experience make a big difference. Professional techniques produce professional results.

In addition to attending workshops and watching videos, another great way to learn the essentials of the trade is to invest in the Professional Decal Application Assn. (PDAA) training course. The five-day class — which provides a tremendous amount of hands-on experience — costs $3,500. For more details on this course, visit the PDAA Website, www.pdaa.com.

Learning from experts improves your installation time, eliminates costly mistakes and helps you create more attractive installations. More importantly, your work will be more profitable and competitively priced.

Job preparations

Before committing to a job, always conduct a detailed survey. Consider the number of vehicles, the substrates' type and condition, location and availability of shop facilities. The information that you gather will help you do a better job of estimating and planning.

Winter weather makes fleet-graphics jobs challenging, and even impossible, in some cases. Tractors and trailers should be washed the day prior to installation. Ideally, the unit should be pulled indoors the night before to sufficiently heat it. This step is absolutely critical if you're installing graphics on heavy construction equipment when temperatures are sub-zero. It can take days for such equipment to reach an adequate installation temperature.

In the real world, however, fleets are often left outdoors until the time for graphics application. Snow and ice that have accumulated on the equipment should be removed before the units are pulled indoors. This prevents water dripping on you and your work as the vehicle defrosts. If water has accumulated on the trailer's roof, lower the vehi-

Photo courtesy of R Tape

When transferring graphics, peel away the liner rather than trying to pull the vinyl away from the liner.

cle's landing gear at the front of the trailer. This creates a slope on the roof, speeding water runoff.

Once indoors, open all the unit's doors to release cold air. Directing portable heaters inside will speed the warming process. Portable heaters can make graphic applications possible, even when the installation facility is inadequately heated. Remember that the application temperature refers to the substrate's temperature, not the ambient air temperature.

Some vinyls can be installed at temperatures as low as 40°F. By misting the adhesive with isopropyl alcohol, you may gain

It's important to cut vinyl along the panel seams. Unslit graphics will inevitably tear because panels move constantly while the truck is in motion.

10° of application temperature. At lower temperatures, the best advice is to go home, throw a log on the fire and wait for a warmer day.

Hot-weather installations can be much more difficult than cold conditions. As the temperature rises, the tack, or stickiness of the adhesive, increases. As the film softens with increased heat, it can distort when repositioned.

We can't change the weather, but we can improve installation conditions. Opening vehicle doors allows heat to escape. Scheduling an installation for an evening or an early morning — or simply waiting for the position of the sun to change — can help you beat the heat.

For smooth surfaces, try misting the surface with a garden hose. This cooling effect can reduce the surface temperature and the adhesive tack to allow for repositioning. Application fluid is also an option. While both techniques work well on flat surfaces, they're not recommended for rivet applications. Any fluid trapped underneath rivet heads usually results in "tenting."

When scheduling jobs, arrange for the vehicle to be washed the day before installation. This allows enough time for the unit to dry completely. Moisture under rivet heads and panel seams can cause problems such as tenting or lifting.

Washing won't remove all surface contamination, so you will also need to prep the substrate again just prior to application using a solvent such as isopropyl alcohol. Saturate a clean rag with solvent and scrub the area receiving the graphics. Before the solvent dries, use clean paper towels to wipe away dirt.

After cleaning the surface, moisture may remain under rivet heads and panel seams. By passing a propane torch's flame over these areas, any residual moisture will bubble and evaporate. Any dirt trapped under rivets or seams will be expelled.

Installation procedure

Carefully study your installation diagram. When laying out a job with overlapping sections, keep in mind that the vertical seams of graphic panels must be "wind-lapped," with the front panel overlapping the rear. Horizontal overlaps must be "rain-lapped," with the top section overlapping the bottom. Each overlap should measure 1⁄4-1⁄2 in.

When installing vertical panels, work from the rear to the front. With horizontal, overlapping panels, apply the bottom section first.

Using a level line on the body, measure the location of each graphic element. When installing trailer graphics, use the bottom rail as a reference point. Using small pieces of masking tape, tape the panels into position and recheck your measurements. Taping all of the graphics sections in place ensures that all elements fit properly within the given space. After a panel is taped in place, use a felt-tip pen or Stabilo pencil to draw registration marks at the graphic's edge from the application paper to the vehicle surface.

When installing small to midsize graphics, a hinging technique is generally unnecessary. However, for extra-long graphics, use a center hinge to make the installation easier. To form a center hinge, apply several pieces of 2-in. masking tape over the premasked graphics. Then, remove the liner paper from half of the graphic, cutting or tearing the liner near the hinge. Squeegee the graphic, beginning near the tape hinge and working toward the outer edge. After completing the first half, remove the hinge and remaining liner paper. Complete the application, starting at the initial squeegee stroke and working outward.

Here's a top-hinge technique for applying tall letters or solid panels that are taller than they are wide. Start by applying 2-in. masking tape along the top edge. For long letters, use scissors to cut between the large individual letters and install each section individually.

When transferring graphics from the release liner, peel away the liner rather than trying to pull the vinyl off the liner. If you're installing a large panel, you may only want to remove a portion, exposing just enough adhesive to apply the vinyl confidently.

Next, align registration marks. Pulling the graphic taut at the top corners, tack it into position without stretching. Stretching the graphic can bow it or create wrinkles that won't come out. Tacking the bottom corners is a matter of personal preference.

To squeegee most graphics, begin in the center and stroke the vinyl with an up-and-down motion. On the "up" stroke, your thumb should be at the bottom of the squeegee. Going downward, pull with your fingers on the top. This first stroke should serve as your centerline with all subsequent strokes following this point. Each stroke must overlap the previous one. Keep the angle between the face of the squeegee and the application

Hingst recommends pulling away application tape at a 180° angle. Tape that's difficult to remove can often be taken off with a light spray of application fluid.

surface shallow as you work. The squeegee should be angled slightly away from the centerline.

Imagine that you're pushing air out from under the vinyl. Never angle the squeegee toward the centerline — this produces bubbles. Maintain good pressure to force all of the air and application fluid from under the vinyl. Squeegee the vinyl over a riveted surface as you would on a flat surface. As you work the material over the rivets, be sure your squeegee moves over the entire rivet.

If you run into a problem such as a wrinkle or bubble as you squeegee the vinyl onto the surface, you'll need to reposition the graphic. This involves jerking some of the material off the substrate with a quick snap of your wrist. Very similar to setting a hook in fishing, this movement should be perpendicular to the substrate. In repositioning the graphic, don't pull the graphic slowly from the surface. Even if you're careful, all you will do is stretch the material out of shape. Once the material is stretched, it's stretched for good.

Before removing application paper, cut the vinyl along the panels' seams. Graphics that aren't slit will eventually tear from the constant movement that occurs between the panels as a truck rumbles down the road.

After cutting the vinyl, use your squeegee's edge to tuck the vinyl into the panel seam. To speed the process, most decal installers prefer to remove the application tape before working the rivets. When slitting the graphics at the seams of roll-up truck doors, the vinyl must be double-cut at a 45° angle along both edges of the joining door panels. All cuts must be resqueegeed to ensure proper adhesion. For roll-up doors, the vinyl should be edge-sealed at the panel seams.

More vinyl maneuvers
When removing the application tape from the graphic, pull the tape against itself at a 180° angle. Stubborn application tape can be more easily removed with application fluid. Lightly spray the tape with application fluid, wait about 30 seconds, and remove it. The fluid penetrates the paper and softens the tape's adhesive, causing the adhesive to release from the graphic.

After removing the application paper, resqueegee the entire graphic, especially all edges and overlaps, to prevent edge lifting. To prevent a hard squeegee from scratching the graphic, use a low-friction sleeve.

After the material is applied, puncture the vinyl in several places around each rivet head using a pin or an air-release tool. Puncturing the vinyl allows air to escape while burnishing around the rivet head. It's faster to use a multiple-pin air release tool. Some installers prefer using a chiropractor's Wartenberg pinwheel, rolling it on either side of the rivet row to puncture the vinyl. Don't use a knife, which creates slashes in the vinyl that eventually open up. A pinhole, on the other hand, self-seals.

There are two tools that make vinyl conform to rivet heads — a squeegee and a rivet brush. In most cases, a rivet brush is preferred. Holding the rivet brush in your fist, move the brush back and forth horizontally over the rivet head with short, choppy, firm strokes. These short strokes compress a larger bubble around the head to a much smaller bubble.

Heating vinyl with a heat gun or a propane torch breaks down the memory of the vinyl so that it permanently conforms to the rivet. A hair dryer won't generate enough heat. If you don't sufficiently heat the film, it will eventually tent around the rivet head. The heating process only takes a couple of seconds. Keep the flame moving to avoid burning or melting the material.

After heating, burnish the vinyl to the rivet head using a circular motion with the rivet brush. This circular motion will spiral inward toward the rivet. Firm pressure with the rivet brush will ensure good adhesion. Rotate your wrist while burnishing, bringing the bristles in at a 45° angle to the edge of the rivet head.

For certain jobs, edges should be coated with a commercial edge sealer. Using a fine-tipped brush, paint the edge sealer on the perimeter of the graphic. Never paint vinyl graphics with any varnish. Varnishes can contain very hot solvents that can penetrate the vinyl facestock and attack the adhesive.

Finally, check if any film has lifted at the edges and that all materials conform tightly to the rivet heads. And, as your mother may have told you, pick up your mess before you leave.

Chapter 34
Applications to Corrugations

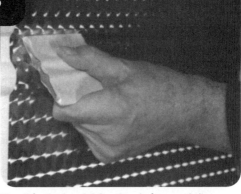

Often, there are several ways to accomplish a task. Applying vinyl to a corrugated trailer surface is no exception. The first installation technique I learned involved using a rivet brush to apply vinyl. I used this technique for years until an installer from California offered his two cents.

It's all in the hands! Good technique and these helpful hints are all you need to successfully apply vinyl over corrugated surfaces.

He believed that using a squeegee rather than a rivet brush stretched the vinyl less, and with less mechanical stress, the vinyl was less likely to tent in corrugation valleys. This technique has served me well for years.

However, approximately 10 years ago, I heard that European installers preferred to apply vinyl to corrugated trailers using hard, felt squeegees. On my next trip to Europe, I visited a Dutch distributor, who gave me a few felt squeegees to try. Guess what? They worked too.

It's a good idea to learn new tricks of the trade. You can learn correct application techniques at the five-day Professional Decal Application Assn. (PDAA) training course (www.pdaa.com). 3M also sponsors training through the United States Graphics Assn. A veteran graphics professional could also be a valuable resource if you can find one willing to train you.

Application techniques' effectiveness will vary. The key is finding a technique that works for you. When you do, continue to use it.

Careful with shortcuts

Not all so-called "tricks of the trade" work. Along with several right ways, wrong ways also exist. Approximately 15 years ago, a colleague called me after having inspected a fleet of corrugated trailers in a Chicago suburb.

He said, "You won't believe what I saw." Before he had a chance to tell me, I interrupted.

"Let me guess," I said. "All the vinyl has lifted in the corrugation valleys, and much of it has cracked and disintegrated."

How did I know? I hadn't seen the graphics fail myself, and I'm not a clairvoyant from the Psychic's Network. But I had seen the graphics' application, and I envisioned future problems. An installer from Chicago's South Side had discovered a shortcut that allowed him to apply vinyl to a corrugated surface quickly with relatively few wrinkles and bubbles.

The shortcut involved tacking the graphic into position at the corners, allowing the film to bridge all corrugations. After heating the vinyl with his torch, the installer used a rivet brush to force the film into the valleys.

However, this presents a problem — vinyl will only stretch so far. Overstretching puts too much mechanical stress on the film. Regardless of your vinyl, this shortcut is a ticking bomb that eventually leads to disastrous results. Sooner or later, the vinyl will lift or tent in corrugation valleys.

How to Apply Vinyl Over Corrugated Surfaces

1. The surface temperature should be at least 50° F. In cold weather, you may need to preheat the trailer surface by using either an additional propane torch or placing a torpedo heater inside the rear door, so it heats from the inside out. Heating requires caution: Excessive heat can cause insulation panels to buckle, causing the aluminum skin to separate.

2. Starting in the emblem's center, squeegee the film to the corrugation's crown. For most applications, professional installers prefer hard, nylon squeegees, such as 3M's gold squeegees. However, softer plastic squeegees, such as 3M's blue squeegees, are preferred for corrugated applications. Softer squeegees will bend and conform around the curved ridges or crowns.

3. Use the squeegee's corner to apply the vinyl to the lower slope of the corrugation. Some manufacturers recommend first applying the vinyl with your thumb prior to using a squeegee. Using a felt squeegee might also work.

4. Run the squeegee's edge in the lower corrugation valley.

5. Starting in the lower valley's center, use overlapping downward strokes to squeegee the film to the flat area between valleys.

6. Use the squeegee's edge to apply vinyl to the upper valley.

7. Use the squeegee's corner to apply the film to the upper slope.

8. Repeat the procedure described in steps 1 through 6.

9. Before removing application tape, cut the vinyl at the seams with a utility knife. Uncut vinyl will inevitably tear as trailer panels vibrate and flex with the trailer's movement. Rather than tearing in a nice straight line, the vinyl rips, leaving ragged, unsightly edges.

Survey your fleet

Before applying vinyl to corrugations, take time to conduct a survey.

Historically, vinyl applied to stainless-steel corrugated trailers has been problematic. Applied graphics frequently fail on these trailers, most often, the vinyl tents around rivet heads. Vinyl should be cut around rivet heads with a rivet cutting tool.

When planning your production, estimate shrinkage of the graphic's height, as the vinyl wraps around the corrugation. Shrinkage varies with the corrugation type. The simplest solution is to apply a 12-in. piece of vinyl to a corrugated trailer surface, and then measure its actual height.

In many cases, trailer surfaces within a fleet will differ, perhaps being a mixture of flat, riveted-flat and various corrugated surfaces. In these cases, designing one size to fit all works best.

If you're completely covering trailer sides with graphics panels, make the graphics large enough to cover the trailer surface and largest shrinkage allotment. Create layouts that allow for trimming without losing critical design elements. This way, the graphics can work with little or no shrinkage.

Choosing materials

Research your vinyl choice for corrugated-surface applications. Study the manufacturers' literature to locate recommended films.

Look for a film with a track record. I have nothing against calendered films or their

10. After cutting the seams, remove the application tape by pulling it 180° against itself.

11. Check for air bubbles under the vinyl. Puncture a bubble edge with an air-release tool or pin and force the air out with your thumb. Never puncture the vinyl with a knife — a knife creates a slash that will eventually open. One or two holes should be perforated at the rivet's bottom. This eliminates water seeping between the substrate and vinyl.

12. Using an industrial heat gun or propane torch, heat the film and run the edge of a squeegee in the corrugation valleys. To prevent the squeegee from scratching or marring the vinyl surface, place it in a low-friction sleeve. An alternative to a low-friction sleeve — which can scratch the surface — is to use the soft side of a Velcro® sheet or strip. Available in strips or rolls, Velcro can be cut to the squeegee's size and applied to one edge. This creates one soft edge and one sharp edge.

13. Puncture the film in several places around the rivet heads with an air-release tool or pin. After heating the vinyl, use a rivet brush in a circular motion to burnish the vinyl around the rivet heads. Whatever application technique or tool you use, successful graphics applications to rivets and corrugations require heat and good mechanical pressure. Heat is important in these applications because it softens the vinyl, letting it conform to the irregular shapes of the corrugations and rivet heads. Heat also breaks the "memory" of cast vinyl films, preventing tenting around rivet heads and channeling in corrugation valleys. Heat also softens the adhesive, making it flow and creating a long-lasting bond.

14. Resqueegee all graphic edges.

15. Inspect your job. Make sure the vinyl conforms tightly to rivets and corrugations.

16. Last, but not least, pick up your mess.

manufacturers, but, in my opinion, cast vinyl is generally the best choice because it's much more conformable (see ST, March 2003, page 44).

You'll also want a film that allows graphic repositioning during application. Films with repositionable adhesives include Avery Graphics' (Painesville, OH), EZ Apply™, Arlon's (Santa Ana, CA) ProFleet™ and 3M Commercial Graphics' (St. Paul, MN) Controltac™ and Comply™.

Prepping the substrate
Arrange for the trailer to be washed the day before installation. This allows enough time for the trailer to dry. Any moisture trapped under rivet heads or overlapping panels will cause the vinyl to tent or lift. Wet applications are strictly verboten.

Washing alone won't remove all contaminants. The next day, remove any residual tar, soot or grease with a cleaner such as Du Pont's™ 3919S PrepSol. The last step involves wiping the surface down using isopropyl alcohol (aka IPA, which isn't to be confused with India Pale Ale).

Before the solvent evaporates, wipe the substrate dry using lint-free paper towels. For complete details on substrate preparation, read the chapter on surface preparation.

Apply yourself
After cleaning the substrate, position the graphic so the top edge lays in the middle of the flat area between corrugations. Begin the installation procedure by peeling approximately

2 ft. of the release liner from the marking to expose the adhesive system. Tack the graphic into position — begin at least 6 in. from the top of the marking. Some installers start even further down, perhaps one-third below the top. When applying circular emblems, some start halfway down.

Whether you use a rivet brush or squeegee, systematically work the corrugation's individual parts. Starting on the crown (also called the ridge), the highest point of the corrugation, imagine looking at a surface cross-section.

Corrugated truck surfaces vary, including the spacing arrangements. Whether a corrugation comprises a rounded-bead or V-shaped design, the basic parts remain: two slopes, two valleys, the crown and usually a flat space that separates the corrugations.

Begin the application on the top of the crown, in the panel's center. Begin the installation sequence by applying the emblem to the crown. Next, affix vinyl to the lower slope, then the lower valley, followed by adhering it to the flat space, then the upper valley, and finishing with the upper slope. For a more detailed step-by-step, review the installation instructions in the sidebar.

Always start each squeegee stroke in the panel's center, applying pressure from the center to the outer edge. Resist the natural tendency to apply strokes in one direction rather than using alternating strokes. This causes alignment problems, especially when one graphic panel must match another. By applying strokes in one direction, you'll likely push the emblem in that direction, distorting the panel.

Chapter 35
Vehicle Wraps

Many signmakers avoid vehicle wraps like the plague — and with good reason. These applications challenge the skills and patience of even the most experienced installers. If you haven't tried one of these applications, I recommend getting some training from an experienced fabricator.

In August 2005, I took a trip for that purpose. I spent two days learning about

When laying out a partial vehicle wrap, use a smooth, flat surface. Areas with numerous rivets and corrugations present challenges to vinyl adherence.

vehicle wraps from experts at cast-film manufacturers Arlon (Santa Ana, CA) and service provider Straight Line Graphics (Huntington Beach, CA). Read on for tips and tricks that should make these applications less challenging.

Why be a wrapper?
Vehicle wraps can be time-consuming, difficult and expensive. So, why produce them? Because a full vehicle wrap is almost impossible to ignore, and, when done properly, becomes effective fleet advertising. When operated in a large city, each wrapped vehicle is seen by millions of people. Compared to other, similar forms of advertising, such as outdoor billboards or transit advertising, the cost per visual impression can be significantly less.

If you use the right film, vinyl vehicle wraps are less expensive to apply and easier to remove than painted graphics. Full wraps also protect the paint system from dents and scratches caused by flying rocks and other road debris. If the vehicle is totally covered, simply remove the vinyl, and, voilá, the paint finish is as good as new with no telltale image ghosting that occurs occurred after having applied computer-cut lettering and symbols.

Design considerations
Some design rules for outdoor advertising and transit advertising are also relevant for designing vehicle wraps. The cardinal rule is to keep the design simple. Customers tend to want to cover every inch of the vehicle with numerous different advertising messages. Your job is to convince him otherwise. Limit the words to a maximum of six. Use a legible typeface, and incorporate bright or contrasting colors to attract attention.

"If the fleet owner is providing the artwork, specify the software, file extensions and the resolution required to blow up a design to a full-size print in terms of dpi," Chuck Bules, Arlon's technical-services manager, said.

When determining vehicle-copy position, avoid such obstructions as door hinges, indentations, handles and compound curves, if possible. It's best to locate text on such flat surfaces as the vehicle's sides, hood and trunk. If you stretch the copy across a flaring, front-quarter panel or bumper, the lettering will likely become distorted, which hampers visibility. In many cases, it's easier to apply cut or printed lettering as a separate step after wrapping is complete. Avoid busy backgrounds that can distract the viewer's attention from the primary message and design elements.

"It's much better to select lighter colors for backgrounds, such as a picture of clouds, sandy beaches or ocean water," Bules says.

Use a pushing motion to direct the squeegee. Keep the squeegee at a low angle near the surface and use short strokes to avoid producing wrinkles and bubbles.

"Fuzzy backgrounds with less defined points of alignment are much easier to match when joining graphic panels. You have much less room for error when you join panels with deep colors and highly defined images."

Tools

Vinyl-application success requires a combination of the right tools, materials and application technique. For most trailer applications, professional installers prefer hard, nylon squeegees, rivet brushes and a propane torch. Vehicle wraps are more easily performed with different tools.

"After a few vehicle wraps, I realized that my trusty gold squeegee wasn't the best tool for the job. It tended to scratch the film, and it wasn't flexible enough when I was working the film into tight corners and around compound curves," Bules said.

Felt and Teflon® fluoroplastic squeegees are now his tools of choice. Felt squeegees are available in various densities and in rectangular and semicircular shapes. Dense, felt squeegees with a sharp edge provide the pressure you need when stretching a film into an indentation or crease on the vehicle's side.

Bules cautions that felt squeegees are only effective for dry applications. Wet felt disintegrates quickly. Bules also prefers the "kinder, gentler" Teflon squeegees (he is, after all, a Californian). The Teflon squeegee's slick surface glides smoothly over the vinyl without scratching it. Its stiff edge provides great pressure, which helps when pushing out hard wrinkles and edge puckers.

Most professional installers prefer propane torches. For a vehicle wrap, a heat gun that generates temperatures up to 400° F is much safer. The less-intense heat diminishes the chance that you'll damage the graphic.

Other practical implements for your toolbox include an air-release tool, a tape measure, 1- and 2-in. masking tape to tack the graphics in place while laying out the job, an Olfa® knife with breakaway blades, paper towels, solvent surface cleaner and denatured alcohol.

Material selection

When installing graphics onto compound curves, you'll need to stretch the vinyl. Problems can start here because vinyl naturally tends to revert to its original shape.

"After several installations, you'll find that you can stretch the film to more than three times its original length," said Bules. "Just because you can, doesn't mean you should."

A vehicle wrap can be very demanding because installers often stretch vinyl to its limits. When a cast vinyl, for example, is stretched 200%, a 2-mil film with a 1-mil adhesive thins to a 1-mil film with a 0.5-mil adhesive.

The same holds true for the ink. When ink thins out it becomes transparent and the color washes out. For this reason, the vinyl you choose must exhibit outstanding performance characteristics.

In the United States, many professional installers prefer cast films with cast-vinyl overlaminates. Cast films are generally very conformable and have excellent durability. Let the buyer beware, though — all cast films aren't the same. Research your material options. Find out what other sign people use.

Calendered films are another option — today's films are more conformable and less bulky than older-generation media. Signmakers usually use them in Europe and other parts of the world.

The signmaker must determine the best vinyl for a shop's needs. If in doubt, conduct your own tests on some cast and calendered samples. Just remember, you usually get what you pay for. Trying to save a few dollars by using a cheaper vinyl will usually cost you in the long run.

Printing and overlaminates

All printed graphics used for vehicle wraps should be protected with an overlaminate or a clearcoat.

"Before applying an overlaminate or clearcoat, make sure that the print is completely dry," Laura Wilson, Roland DGA Corporation's (Irvine, CA) product manager said. "If you clearcoat, it's better to apply two thin coats instead of one thick coat. This prevents trapping any residual solvents."

Many companies that specialize in vehicle wraps prefer overlaminates because they give the film extra rigidity. "We use cast-vinyl film from Arlon with their cast-vinyl overlaminate," John Checkal, Straight Line Graphics' (Huntington Beach, CA) owner, said. "It has high adhesion and seems to handle a lot of distortion without losing tack or adhesion."

When laminating the print, don't

An initial, horizontal squeegee stroke tacks the graphic into place. Subsequent squeegee strokes across the graphic should overlap each other. Trimming excess vinyl will relieve some of the graphic's tension and make the film easier to apply.

use too much unwind brake pressure or the overlaminate can stretch. Checkal said after the graphic has just been printed, the film and adhesive are in their softest state because of some residual solvents. Excessive laminate tension triggers graphic tension, which causes the entire finished product to shrink. Within a day or two, while the graphic remains on its release liner, shrinkage will occur at the edges. A few months after installation, the edges will start to shrink and lift at the overlaps.

For most applications, the nylon squeegee is the professional's tool of choice. However, for vehicle wraps, a solid Teflon® fluoropolymer-resin squeegee works better because it slides across the vinyl surface with less drag.

Checkal also recommended using an ultra-low-tack premask. He claimed the protective masking gives the print extra stiffness, which eases graphic application and saves time.

"When you apply graphics outside in the heat, cast vinyls tend to stretch," he said. "The masking holds everything together so one panel lines up perfectly with another with no misregistration."

Being efficient

Installations are challenging because they impose many obstructions, such as hinges, gas-cap doors and compound curves. Chuck Bules, Arlon's technical-services manager, recommends removing mirrors, lights and any other detachable impediments.

"If you don't remove these obstructions, you can waste a lot of time cutting, tucking and wrapping the film," he said. "Usually, it's much more efficient to remove the obstructions, apply the film and replace the parts. And, organize all of the parts in one place."

By taking these extra steps, installations are usually completed faster and look more professional.

Surface preparation

Successful installation of any vinyl graphics depends on proper surface preparation. Dirty surfaces and poor application techniques most often cause adhesion failure.

Surface preparation is a three-step process. The difference between installing a logo and a few words on the truck's side and completely covering a vehicle with graphics is that vehicle wraps require cleaning every part of the vehicle. This includes the paint at the bottom of the wheel wells, inside the doors and door jambs, and all around the gas-cap door.

Make sure the vehicle is spotless before you start. First, wash the surface with detergent and water. Second, use a solvent cleaner to remove grease, tar and waxes. Finally, wipe the surface with an alcohol-and-water mixture.

Even though a washed surface looks clean, such contaminants as waxes, grease and oils are most likely still present. To remove these contaminants, use DuPont®'s 39192 Prep-Sol or a grease and wax remover. In the sign trade, many graphics installers use such products as Rapid Prep or Universal Products' TFX Professional Striping Cleaner.

After you use the solvent surface cleaner, give the surface a final wipe with denatured or isopropyl alcohol. This last step should be performed with both hands. In one hand, you'll need a rag saturated with alcohol. In your other hand, keep a clean rag or paper towel. After you apply the solvent, wipe it dry before it evaporates.

Use caution — as I always say, test, don't guess. To ensure you don't damage the paint system of a car or truck, test the solvent on an inconspicuous area of the vehicle before using it. Doing the final cleaning right before vinyl application removes airborne contaminants that may have settled on the vehicle's surface.

Wet or dry?

Wouldn't it be nice if you could perform every application in a climate-controlled bubble? Unfortunately, life doesn't work that way. Sometimes, we have to contend with summertime heat, which may cause adhesives to stick prematurely. When working in hot weather, many people resort to application fluid.

If you can complete the application dry, do so. Many, though, will opt for a wet application. If that's your choice, here are a few suggestions. First, use a commercial-grade application fluid such as RapidTac rather than a homemade concoction. Commercial-grade products ensure consistency.

Also, use the least amount of application fluid necessary. For wet applications, use good squeegee pressure to expel all fluid from underneath the vinyl graphics. Any residual fluid contaminates the graphic's adhesive, which can lead to adhesion failure.

"Application fluid can impair speed if you use too much of it. If you're in a real hurry, apply the film dry," Bules said. "Never use application fluid as a crutch. It's better to learn good installation techniques and continually practice your craft to hone your skills."

The installation process

When handling large sections of sticky graphics, you may have wished that God had given us two extra hands.

For vehicle wraps, cut the installation time in less than half by working with a partner. An extra set of hands can help hold the graphic away from the application surface and prevent pre-adhesion accidents. And a partner can help stretch the material to conform around compound curves.

Before starting the application, tape all of the pieces in place to ensure they reach the correct spot. This is also the time to plan the installation sequence. Many installers start with the hood, and then install the sides. Other installers start from the rear of the vehicle and progress towards the front. Decide what works for you.

Chapter 36
Wet Applications

During recent travels to meets and shops, I've spoken with several signmakers who've posed questions and problems relating to wet vinyl applications. Although I've never been a big advocate of wet applications, I realize that many graphics professionals prefer them. Thus, I'd like to offer a few tips to help you perform them more effectively.

Five easy pieces (of advice)

If you can install vinyl dry, never use application fluid. Wet applications should only be used as a last resort. When I learned how to install fleet markings, I applied everything dry. If you've perfected your technique and you're using vinyl with a forgiving adhesive system, dry applications are the faster way to go.

As part of my indoctrination, my instructor warned me not to use application fluid to apply vinyl to surfaces with rivets or reflective sheeting to unpainted metal surfaces because application fluid collects underneath rivet heads, creating a residue that later seeps out and causes vinyl to tent and eventually crack. Fluid under trailer-panel seams can also cause edge lifting.

Using application fluid to install graphics with a metallization layer, such as reflective sheeting can cause delamination of the face-stock's adhesive and accelerate galvanic corrosion. When two dissimilar metals — such as aluminum in reflective sheeting and the skin of an unpainted stainless-steel tanker truck — come in contact, the reaction generates an electric current. Electrical particles (ions) flow from the reflective material's metallized layer into the steel.

During an electrical reaction, one becomes the donor metal, while the other becomes the acceptor metal. As the aluminum in the reflective sheeting loses ions, it corrodes and blackens. By acting as a conductor between the metals, application fluid can facilitate the electric flow of ions, which leads to corrosion.

Second, always use application fluid to adhere vinyl to an acrylic sign face. Here, an application fluid is required because the vinyl's adhesive wants to grab onto the plastic surface.

I once tried to install vinyl via dry application to a plastic sign face. The result: a graphic with a zillion tiny bubbles. Although all these bubbles eventually "exhaled" and disappeared after three days of summer heat, I learned the correct way to apply vinyl graphics to a flat plastic surface: Use a commercial application fluid.

Note that I recommend a "commercial" product, which leads to my third piece of advice. Years ago, I told signmakers they could make their own concoction by mixing 20 oz. of water with 1/2 tsp. of a dishwashing liquid and 1/2 tsp. of isopropyl alcohol.

At that time, I believed application fluid was that simple. I was wrong. In doing a side-by-side test of my mixture vs. the real stuff, I learned that commercial fluids work better because they promote faster vinyl adhesion to the substrate.

Some application fluids cost as much as $25 a gallon, which prompts some signmakers to "extend" their supply of the mixture by adding water. Remember, if you dilute your application fluid, you'll dilute the adhesion-promoting characteristics.

Commercial application fluids are worth the investment because the formulations' quality and consistency yield consistent results. Some popular products on the market today include Rapid Tac, Rapid Tac II, Splash, Actiontac, Window Juice, Quick Stick and Position Perfect.

Don't waste your time trying to duplicate these products because you'll never achieve the same consistency. Dishwashing liquid and similar soaps contain additives such as surfactants, emulsifiers, moisturizers and perfumes, all of which are detrimental to an adhesive. Surfactants, for example, help cleaners break dirt's bond with the substrate. They have the same effect on adhesives, causing bonding failure and edge lifting.

An application fluid aids installations if you're applying a vinyl with an aggressive adhesive system. The application fluid helps float the graphic onto the surface to prevent pre-adhesion (the vinyl sticking before you want it to). Application fluid also allows an installer to easily reposition the graphics without distortion.

My first wet-application attempt occurred 20 years ago while installing "prototype" window graphics for a chain of hardware stores in Louisville, KY. The hot, sunny weather caused the vinyl's adhesive to stick to the glass prematurely.

These conditions warranted using an application fluid. Weeks earlier, our sales vice president had recommended using Windex® glass cleaner as an application fluid, so we gave it a try. After the installation, as we said good-bye to the chain's general manager, I looked over my shoulder at the windows. To my dismay, all the graphics had slid down the glass. Luckily, because this was a prototype, we'd produced twice as much material as we needed. The temperature had cooled considerably by this time, and we dry-installed the graphics.

So, my fourth tip is: Never use glass cleaner as an application fluid. Some glass cleaners contain silicone, ammonia and other additives that can impair adhesion (Windex isn't even

At a recent Chicago seminar, Jim demonstrated how to properly complete a wet application. (From left) He squeegeed the graphic with intact application tape, removed the tape and squeegeed once more after the tape's removal.

recommended for preparing the application surface).

My final point: Disregard technical advice from your VP of sales. If you want technical information, talk to a technical service person.

Step-by-step instructions
- Cleanliness is next to godliness. Before any application, clean the substrate thoroughly with a commercial, non-abrasive cleaner. A final cleaning with isopropyl alcohol is generally recommended. When wiping a surface clean, use a lint-free rag or paper towel.

- When performing wet applications — or any application — select a low-tack application tape. Generally, I recommend using a lower-tack tape for large lettering and graphics and a high-tack tape for smaller lettering and finer detail. In choosing the right application tape, my advice is, "Test, don't guess."

- When using application fluids, less is usually more. Lightly mist the substrate with application fluid, using the least amount necessary to accomplish the job. Don't spray the adhesive side of the vinyl graphics and application tape. This often causes the graphics to prematurely release from the tape, resulting in installation mishaps.
  Many signmakers use much more fluid than is needed. At one signshop, the installer used so much application fluid that he wore rubber boots so his feet wouldn't get wet. Using that much application fluid is ridiculous.

- Use a hard, nylon squeegee with firm pressure and overlapping strokes, starting in the center and working outward to force out the fluid from underneath the graphics. Remember, you're applying a pressure-sensitive film. It's called "pressure-sensitive" for a reason, so apply some pressure.

- After a few minutes, remove the application tape by carefully pulling it 180° against itself. To expedite the process when using paper tapes, spray the tape's backside with application fluid. The fluid will penetrate the paper and soften the adhesive, allowing the tape to release more quickly.
  Using too much application fluid and failing to remove the application tape promptly can cause the adhesive to delaminate from the tape. Cleaning this residual adhesive can be annoying and time-consuming.

- After removing the application tape, mop up the residue with a paper towel.

- After the application tape is removed, use a squeegee covered with a low-friction sleeve and squeegee the entire graphic again. The low-friction sleeve prevents the squeegee from scratching the bare vinyl. This step prevents edge lifting and ensures good adhesion.

Chapter 37
Application FAQs

Recently I traveled through Europe and South America, speaking to groups of signmakers about vinyl applications. Because my talks cover only the basics for applying films over flat surfaces, rivets and corrugations, I received several questions about special applications.

Everyone who applies vinyl graphics periodically encounters vinyl installations on windows, drywall and compound curves. Here are a few answers to the most frequently asked questions.

Houston-based Natural Graphics fabricated this apartment-ID sign, which earned second place in ST's Commercial Sign Contest, by painting Gerber vinyl and applying it to Duraply™ plywood with redwood accents.

Automotive pinstriping

Applying automotive pinstriping, which is usually thinner than an inch, requires a delicate touch, as opposed to installing truck graphics. Although many signmakers cut their striping on plotters, prespaced automotive striping and printed-graphics kits are available at automotive warehouse distributors, body shops and sign-supply distributors. Striping colors usually match sign vinyl.

To install pinstriping in a straight line, first remove 2 ft. of the release liner. Tack one end of the striping to the car body, then stretch the remaining portion to the vehicle's opposite end. As you work the stripe into position, remove the remaining liner paper. The ideal alignment usually requires repositioning the striping several times. The car's body lines usually serve as the best reference points for alignment.

When the final positioning is determined, tack the stripe and keep the material taut – it can be slightly stretched. Overstretching the material, however, can bow the striping.

Initially, pat the striping with a cloth folded into a pad rather than squeegeeing the vinyl. At this point, a hard, stiff squeegee tends to stretch, buckle and distort the delicate pinstripe, which causes waves.

Automotive striping is usually masked with a clear application film. Always remove the application pinstriping tape after completing the application. Pinstriping usually isn't wrapped inside the door jamb. Instead, the stripe is cut back 1/4 in. from a door's edge. After placing the pinstriping, you can apply more pressure with either your thumb or a squeegee with a low-friction sleeve.

Some installers prefer to wrap wider vinyl around the door's edges and inside the jamb. Be sure to clean any surface before applying the vinyl.

Installing striping wider than an inch requires a different technique than pinstriping. After you tack the stripe into place, place the edge of your squeegee in the middle of the stripe to prevent it from moving up or down, which would produce a wavy stripe. Squeegee strokes travel from the middle to the stripe's outer edge.

Always overlap your strokes. Never squeegee with a stroke parallel to the length of the stripe. Lengthwise strokes can cause the strip to stretch, buckle or wave. After squeegeeing, cut the stripe at the edge of the body panels. After removing the application tape, always re-squeegee the striping.

Curved surfaces

I'd like to advise you to never stretch any film, but this isn't practical for vehicle applications. Vehicle surfaces usually have twists and turns, and the film must conform to these surfaces.

In many cases, it's best to cut the vinyl around these areas. If you stretch vinyl, you can overstress it, which can cause it to fail. Vinyl is also like a rubber band. If you stretch it, it will try to stretch back to its original shape.

When you're stretching vinyl over a convex shape, into a concave surface or over compound curves, film manipulation is easier if you remove the application tape.

When confronted with indented areas in vans, don't force vinyl into an indentation. Stretching the film, especially when it's cold, can break it. Instead, first apply the vinyl into the indentation's flat area. Then, very carefully press the film into the corners. If you need to stretch the film, apply heat to break the vinyl's memory. Don't overheat or overstretch the graphic or you will damage it.

The success of applying a film to a curved surface depends on the appropriate film selection. Some films are more forgiving. Cast films, for example, stretch more easily than calendered films. An additional overlaminate layer makes a graphic less conformable. Polyester films (plastic), by their very nature, are very rigid and can't be stretched over even simple curves.

Select a film suited to the application, and then apply it without overstressing it. I prefer a cast film with a repositionable adhesive.

When you stretch a film, application temperature becomes more critical. At lower temperatures, films become brittle. If you don't handle the film carefully, it could snap.

If you're installing graphics to a convex surface — an air dam on the top of a tractor — the vinyl may bunch up in the corners. You may need to cut excess material and overlap the film.

When installing graphics to air deflectors atop a tractor cab, you'll usually straddle a ladder to the top of the tractor. When I worked on a project for a leasing company, the air deflectors would arrive separately from the power units. My friends at the leasing company called me when they arrived so I could install them on the ground. This is an unbelievable timesaver.

Key holes and bolts

Eventually, you will be asked to perform unreasonable applications. An easygoing, accommodating person may acquiesce to these requests, even if it's wrong. Don't do it. Either decline the job or ask the customer to sign a statement that releases you from any responsibility. Even then, if the application fails, your customer likely will blame you and refuse to pay his bill.

Don't apply vinyl to door hinges, bolts, high-profile rivets, rubber gaskets and door locks. The film must be cut around — not applied to — these obstructions. Remove obstructing hardware and replace it after the vinyl application. This works better than trimming vinyl around an object.

To cut vinyl around an obstruction, first cut an "x" over the bolt or keyhole. Then, using a squeegee or your thumb, tuck the vinyl in tightly around the obstruction. Finally, trim the vinyl with an X-acto® knife and apply more thumb pressure to ensure that the vinyl edge adheres well.

In some applications — high-profile rivets, rivets spaced too closely together and rivets on stainless-steel trailers — vinyl can't be applied over the rivets. Such films as conspicuity-reflective films won't conform to rivets. To cut vinyl film around

Flyway Signs (Fond du Lac, WI) created graphics for this seafood restaurant by applying an unpainted-aluminum façade to the concrete silo, then applying 3M pressure-sensitive vinyl printed on a Gerber Maxx using process and spot color.

rivet heads, 3M (St. Paul, MN) has developed rivet-cutting tools that you should have. Check with your distributor, or examine the 3M tool catalog. When I did this many years ago, I ordered nearly everything.

The rivet-cutting tools include a thermal die that slips onto the end of a soldering gun. The heated tip melts the vinyl without having to apply any pressure. After you cut the film, remove the material over the rivet head.

The more traditional, rivet-cutting tool, which resembles a hole punch that cuts grommet holes, has a handle that helps twist the tool to cut through the vinyl.

Roll-up doors
Before applying vinyl to roll-up doors, thoroughly clean the doors, especially the seams between the door panels. As you lift up on the door, the seam will open up, which allows you to wash and dry these edges. This prevents peeling.

After you apply the graphic and remove the application tape, double-cut the vinyl at each seam; that is, angle your knife at 45° and cut the film along the top edge of one panel. Along the bottom edge of the adjoining panel, make another cut at another 45° angle. To promote better adhesion, heat the seams with a heat gun or propane torch and then re-squeegee all edges. Then, edge-seal all vinyl at these seams. Edge sealing isn't an option; it's a requirement.

Window-graphics film

Window graphics require special surface preparation. Before installing the graphics, thoroughly clean the window. First, wash the windows with a liquid detergent and water. Then, re-clean the glass with isopropyl alcohol. Don't use such glass cleaners as Windex® that contain silicones and ammonia, which can leave a residue that could hinder adhesion. After washing carefully, inspect the windows. To clean windows speckled with dried paint, use a razor blade and re-wipe the windows with alcohol.

Keep your hands, which are dirt magnets, spotlessly clean when installing graphics to windows. When I install window graphics, I wash my hands in alcohol. It may not be healthy, but it works. Dirty hands transfer the dirt to the film's adhesive side. Many customers notice and object.

If you're applying film to glass in high humidity and temperatures are below the dew point, moisture can condense on the surface, which could inhibit adhesion. Be patient and wait for warmer temperatures.

Window graphics should be applied dry. On a dry window application, squeegee marks may appear on the film's adhesive side after the graphic is installed. Assure your customer this is only temporary. The adhesive will flow out in four or five days, and the squeegee marks will disappear.

The same holds true for tiny bubbles underneath the vinyl. Time and temperature breathe these bubbles out of the film. Don't puncture the little bubbles with a pin or knife. A worse problem is puncturing the film and creating a very noticeable hole through which light can shine.

To a certain extent, application fluid contaminates the adhesive. If a wet application is required, use appropriate squeegee pressure and overlap your strokes to force out any moisture underneath the vinyl. In cold climates, any remaining moisture could freeze and cause adhesion problems.

Usually, when you apply graphics in multiple panels, overlap the vinyl at least 1/4 in., where the panels form a seam. But don't do this to perforated window-graphic films, which comprise thousands of little holes that represent 40 to 50% of the total area. Perforated window-graphic films allow the printed image to be seen outside of a window. However, inside the window, the viewer still can see through it.

Because the film has so many little holes, there's approximately half as much adhesive on the top, overlapping sheet. The bottom graphics panel has roughly half as much surface area to stick to. Instead, butt the panels together. If you overlap the film, edge-lifting problems can also occur where the vinyl touches the edge of the window. Always trim 1/4 in. of the film away from the edge. To minimize edge lifting, you can always edge seal the graphic.

When a colleague and I inspected graphics panels with overlapping seams, we saw that most exhibited edge lift. When graphics start to peel at the edges, trim the peeled material and then edge seal the graphic.

Some vinyl companies recommend or require an overlaminate with their perforated window-graphics film. This precludes using application fluid because fluid would be trapped inside each little hole. In the real world, very few people will use an overlaminate. Without an overlaminate, edges can lift and collect dirt.

Applications to Textured Surfaces

With proper technique, you can apply cast, vinyl film to such textured surfaces as an architectural building panel, banner material with a heavy scrim or concrete.

All vinyl graphics require a few precautions, but nothing guarantees a trouble-free application. Usually, applying vinyl to concrete, cinder block and brick isn't recommended because the surfaces are rough and porous. Instead of installing a graphic on a concrete wall, use a banner.

However, to successfully adhere vinyl to these surfaces, seal the graphics. First, squeegee the vinyl graphic into position. After removing the application tape, heat the film with your heat gun and burnish the vinyl using a rivet brush. With a circular motion, use the rivet brush to work the vinyl into the substrate's texture. The finished graphic should look like paint.

Wood

Can you apply vinyl to wood? Absolutely — if you correctly paint the wood first.

Vinyl shouldn't be applied to any unpainted, porous surface. The substrate will absorb moisture, which will break the adhesive bond. Imperfections on the edges of the wood-panel surface must be filled.

Before painting, sand the sign blank and wipe it clean with a tack cloth. A primer coat seals the wood surface and helps anchor subsequent coatings. After priming, add at least two coats of paint to the surface. The board's backside and edges should also be painted. Light sanding between coats help one layer of paint bond to another.

Unpainted metal

Some people believe vinyl shouldn't be applied to any unpainted metal surface. It's problematic. You'll probably be asked to apply graphics to unpainted, aluminum tankers and stainless-steel trailers. You can either politely refuse these applications or take a few suggestions. Remember, surface preparation impacts adhesion.

Because unpainted aluminum rapidly oxidizes, its surface must be degreased and etched with a commercial, acidic brightener. After etching, clean the surface with a solvent before applying vinyl. You can also mechanically abrade the oxidation using steel wool or a 3M Scotchbrite® pad. After abrading the surface, wipe the metal with a towel saturated in alcohol. Before the solvent evaporates, wipe the surface with a clean, dry towel.

Stainless-steel trailers have a long history of vinyl-film failures. Stainless steel, which is porous and easily collects grease, can contaminate the film's adhesive. Thus, film lifting or tenting around rivet heads may ensue.

Installers have tried many solvents and cleaning techniques. Some have even tried, unsuccessfully, to burn grease off rivets with a propane torch. The best recommendation is to avoid stainless-steel trailers. Failing that, cut the vinyl film around each rivet.

Drywall

You can apply vinyl to walls. Drywall, however, must be primed and painted with a gloss or semi-gloss paint. Latex paints contain such chemicals as surfactants and silicones, which can prevent vinyl adhesion.

Before priming and painting new drywall, sand the wallboard smooth. Wall-surface imperfections will be even more noticeable on an an applied vinyl graphics' glossy surface. Before painting, wipe the surface with a clean rag to remove any powdery residue. The primer seals the surface so the wall's paint coat appears even and also helps bond or anchor subsequent coats. A good paint coat will also hide wall imperfections.

Even if a wall looks clean, wipe it down with a damp rag before installing graphics on it.

Underwater applications

Believe it or not, a few people have asked me about applying vinyl to underwater surfaces. One application involved applying a gigantic logo to the bottom of a stainless-steel swimming pool. Vinyl is just not intended for some applications, and this is one of them.

However, a technical service person for a vinyl company told me of an application that worked. The graphic was applied to the glass bottom of a tour boat in Florida. After the vinyl was installed, the entire emblem was clearcoated with Butch Anton's "Frog Juice." Wonders never cease.

Fuel dispensers

I learned to correctly install graphics on the fuel dispensers' pump skirts when installing store graphics in Michigan during the last Super Bowl weekend. As you can imagine, the temperature outside was too cold to install vinyl. Rather than return another day, we removed the panels from the pumps and took them indoors. Whether the weather is freezing cold or steaming hot, this works best.

It's much easier to install graphics on a worktable than to kneel on the ground. You also avoid the annoyances associated with car traffic at the gas station. Plus, you don't disrupt your customer's business. Most importantly, you can do a much better installation job because you can wrap the vinyl around the panels' edges.

Any hardware or locks on the pump skirt should be removed. Clean the surface where the vinyl will be applied. After applying the vinyl to the skirt's face, cut the film at each corner to form 90° angles, then wrap the film around the edges. If the pump skirt's facet has openings for locks or hardware, cut the vinyl over each opening and fold the vinyl into the opening. Then, reinstall any parts you may have removed.

Chapter 38
Removals

Soon after starting in the fleet- graphics business, I made my first big sale — a $26,000 job that involved removing old trailer graphics and applying new markings. In the language of decal installers, the respective processes are called de-identification and re-identification.

I was proud of my sales prowess, walking tall and strutting my stuff. However, I was soon given a reality check, and learned a little about humility and a lot about removals.

I had bitten off more than I could chew — our company didn't have the manpower to keep up with the work. To keep my

After heating the surface with a "torpedo" furnace, vinyl typically removes easily.

customer happy, I actually exchanged my suit for work clothes at the end of the day and helped scrape off old vinyl.

I also learned something that you may have already discovered — removal jobs are never fun, rarely profitable and frequently plagued with problems. While it's usually best to pass on these jobs, removals are often an unavoidable part of a graphics program.

No bowl of cherries
To date, no one has discovered a secret formula that makes old vinyl immediately disappear from the surface that it covers. Vinyl graphics only peel off without a trace of adhesive residue when they shouldn't. Planned vinyl removals usually require long, frustrating hours of hard work.

Several citrus-based adhesive removers have been developed that considerably reduce the potential hazards to the environment, substrate and product user. Two such removers are Orange Peel, marketed by Graphic Adhesives Products (Burbank, CA), and Rapid Remover from Rapid Tac (Merlin, OR).

New plastic blades allow signmakers to scrape old vinyl adhesives without gouging, cutting or scratching the substrate. And power-drill vinyl erasers quickly abrade both facestock and adhesive.

No product, however, replaces the old-fashioned elbow grease required for almost all removals. Several factors make this chore hard work. As vinyl gets older,

Hardly an emergency, this "911" was also removed from the same police cruiser using a torpedo furnace.

the leaching plasticizer tends to make the film brittle. Picking off little bits and pieces of film is aggravating and time-consuming. During de-identification, the adhesive can also delaminate from the facestock.

To minimize the tedium, signmakers and decal installers have developed various graphics-removal procedures. Most employ a combination of heat, chemicals and tools.

Traditional removal procedures

The most common, reliable method of vinyl removal is to apply heat and then peel the film from the surface. The heating process softens the vinyl facestock and the adhesive. A propane torch or industrial heat gun provides sufficient heat to remove smaller areas containing letters and graphics.

Larger areas, however, need to be heated. A huge propane torch — known as a weed burner — is a perfect heat source for these jobs. With an extra long hose, the propane tank can be left on the ground, safely leaving the scaffolding free of clutter.

Heat a large section of the truck surface for approximately one minute. Keep the flame moving so as not to burn the vinyl or substrate.

To keep the surface hot when removing graphics, you can also direct heat from a torpedo furnace — also known as a "salamander furnace" — onto the truck's inside wall. The furnace heat will keep the surface warm much longer, which helps keep the adhesive and vinyl soft and pliable.

After heating, use a fingernail, a Teflon®-coated plastic scraper or a plastic blade to lift the graphic's edge. When lifting the vinyl from the surface, pull at a low angle close to the work surface.

The secret is to apply the correct degree of heat, something that can only be learned by trial and error. If the vinyl is too hot or too cold, not only will you leave the adhesive, but the film itself will break as you pull it from the surface. Various factors dictate the ease with which the film is removed: the age of the graphic, the substrate's condition and the vinyl type.

Photo courtesy of R Tape

This vinyl is being removed using a chemical process. Remember to start with weaker removers. Move to more toxic chemicals only as a last resort because of their potential long-term health risks.

Graphics from the same roll of vinyl applied to different substrates will remove with various degrees of difficulty. I learned this lesson — a vital one in the correct estimation of removals — the hard way.

Approximately 20 years ago, I quoted on a removal job involving a dozen trailers in Florence, KY. I decided to remove the graphics on half of a trailer, which took less than two hours. Based on that, I projected the job should only take four hours per vehicle to accomplish. My boss advised me to play it safe and price the job at six hours.

Murphy must have had graphics removals in mind when he formulated his law. The job required an average of 12 hours per trailer.

Graphics applied to new trailers with a smooth, factory-paint finish removed easily, while those applied to old, pitted truck bodies were extremely difficult to remove. The rough finish provided the adhesive with a greater total surface area. Consequently, the adhesive bonded better to the substrate than it did to the vinyl. When the vinyl was removed, the adhesive remained.

Hazardous to your health

Adhesive removal requires the use of chemicals. Play it safe and always adhere to the manufacturer's instructions. Many chemical-removal systems pose serious health risks because their toxins often enter the body by inhalation or skin absorption.

If you use dangerous materials, your tool kit should contain the appropriate safety equipment: chemical gloves, air respirator and goggles. Prolonged exposure can cause permanent health complications, including brain damage. Other solvents are carcinogenic, while strong acids and alkalis can cause serious chemical burns.

Before working with an adhesive remover, always test the chemical on an inconspicuous spot of the substrate to make sure the remover doesn't react with the paint.

Repainted vehicle surfaces are especially susceptible to damage from chemical removers. In a few graphic-removing jobs, I've stripped paint right down to either the primer or bare metal. Such accidents are embarrassing, to say the least.

Many different chemicals, including isopropyl alcohol, PrepSol™, kerosene, lacquer thinner, xylene and citrus-based removers, can remove adhesives. To reduce the risk of paint damage and minimize any health hazards, start with a milder remover, such as a citrus-based product.

Spray the remover on the adhesive residue. When the adhesive softens to a jelly-like substance, use a squeegee to scrape the gel from the surface. Old rivet brushes can scrub adhesive off the rivet heads.

Working with chemicals

A few years ago, Mod Industries (Rolling Meadows, IL) introduced XXL 1000 Decal/ Adhesive Remover. The substance contains three mild acids that are reportedly environmentally safe and non-flammable. Users can apply XXL 1000 with a garden sprayer, paintbrush or roller.

Film and adhesive removers penetrate the facestock so that they attack the adhesive system. These removers may not work with every type of film, such as reflectives or graphics protected with a polyester overlaminate.

In about 20 minutes, the film and adhesive soften. When the film starts bubbling, the user can peel the vinyl from the substrate or blow it off using hot water and a power sprayer.

Industrial-grade power washers, with the capacity to generate up to 3,000 psi, cost $1,000 and up. If you don't have a regular need for such equipment, consider renting a system when necessary.

Use caution when handling a high-pressure sprayer — excessive pressure will remove

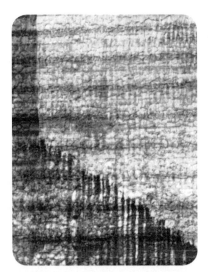

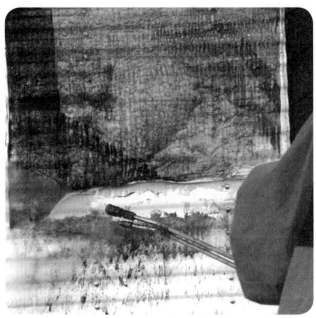

(Above) Power-washing away old vinyl can be a quick process. However, consider the condition of the graphic surface and the substrate when estimating time required for removal. (Right) Don't use excessive pressure when power-washing – otherwise, the vehicle's paint and finish will peel off along with the graphics.

not only the graphics, but potentially the paint finish as well. First, try to remove the film by hand.

If a power sprayer is required, start with a lower pressure — such as 800 psi — and increase pressure as needed. Keeping the nozzle of the sprayer at least nine inches from the surface will also help prevent damage to the finish.

When working with the power sprayer, begin at a vertical edge of the graphic and spray with a steady, up-and-down motion.

In some cases, adhesive residue remains on the surface. When this happens, spray the residue with remover, wait for the adhesive to soften, and wipe the surface clean with rags or paper towels. Finish cleaning the surface by wiping with isopropyl alcohol.

The surface must be perfectly clean before installing new graphics. Applying new vinyl over old adhesive practically guarantees film failure.

During the removal process, the adhesive will absorb the chemicals like a sponge, although some of the chemical will evaporate and the residual adhesive retains the rest. If the new graphics are applied over the residue, the remaining remover will attack the new adhesive. This can cause new vinyl to bubble, peel or fall off.

For baked-on truck finishes and pigmented vinyl graphics, a chemical remover is the solution. For many types of removals, however, chemical vinyl removers are not recommended. Removing screenprinted decals with them can be extremely messy because the remover quickly dissolves the ink and clearcoat, causing the colors to drip on the vehicle.

To prevent staining, first mask the substrate with a premium-grade application tape. Then, apply masking tape to secure the edges protected by the application tape. Use a pressure sprayer to wash off dripping ink immediately.

Film and adhesive removers are often so potent they strip away everything except factory paint finishes. My best advice is to "test, don't guess," when using any chemical-removal system.

Other removal options

During vinyl removal, many installers have experienced injuries resulting from exposure to heat and chemicals. Blistered fingers, dizziness and chemical burns are common maladies.

Using scrapers, plastic razor blades and plastic abrader wheels, cold removal can prevent these hazards. This method frequently works well with calendered films, because they possess much higher tensile strength than cast films and often can be pulled away from the sign substrate in one piece.

Rubber-abrading wheels are ideal for removing small lettering and pinstriping. Abrasive wheels, such as the Stripe Eliminator, bolt into the chuck of an electric power drill. For this work, use a professional-grade drill that can reach at least 1,200 rpm.

The spinning wheel works like an eraser, rubbing off the vinyl and adhesive in one operation. Because of the time required, this isn't usually a suitable method for removing large-format fleet graphics. A word of caution — this tool only works on graphics applied to factory paint finishes, glass and other durable surfaces. Test the abrading wheel first to avoid grinding the paint off the substrate.

Compared to using heat to remove window graphics, the cold method is a safer alternative to prevent glass breakage.

The bottom line

Although nothing can eliminate unforeseen problems, the following tips might make de-identification a more profitable activity:

- If your customer agrees, work on a time-and-materials basis. An arrangement such as this can prevent you from losing your shirt on the job.
- Carefully survey the job before estimating. Look for potential problems, such as graphics applied to old surfaces.
- Attempt to remove some of the graphics before quoting a price.
- Remember to include all the necessary extras in your estimate, such as travel time, equipment rental and the cost of chemical removers.
- Removals usually take longer — a lot longer — than predicted. Always cover yourself by increasing your estimated time by a factor of 1.5 to 2 times.
- Get someone else to do the work for you. If you can, use an experienced, reliable professional decal installer as your subcontractor.
- If you have any inkling that graphics removal will be a problem, it probably will. Let your competitors waste their time with problem accounts.

Chapter 39
Troubleshooting Vinyl Failures

There's nothing more satisfying than solving a good
graphics mystery. That's
usually true if you're a technical service person who
thrives on troubleshooting
challenges and has no financial stake in the job.

This window graphic was made from calendered vinyl, which is
prone to cracking and is often a poor choice for vinyl that will be
subjected to rough conditions.

But in the real world of the
signmaker, when graphics
fail, someone pays for it. If
that failure affects your pocketbook, that enjoyable intellectual exercise often becomes
emotionally charged, and there's nothing satisfying about that.

From the signmaker's perspective, what's really important about troubleshooting is
discovering the cause of the failure so it's not repeated.

Common-sense suggestions
As with most things in life, there are right and wrong ways to handle complaints. Here
are a few of my gems of wisdom on the subject.

Respond quickly. One very wrong way to handle a problem is to ignore it, hoping that it
will just go away. Years ago, I worked with a classic procrastinator who documented
complaints in detail. There's nothing wrong with that, but these documents were just filed
in his desk's bottom drawer with no action taken. After three years of inaction, more than
400 unresolved complaints had accumulated. I had the misfortune of cleaning up my co-
worker's mess, which took several months.

None of these problems went away; they just festered in customers' minds. The longer a
complaint drags on, the more difficult it is to resolve. In my opinion, when there's a
problem, deal with it now. At the very least, visit the complaining customer promptly to
demonstrate your concern.

In troubleshooting the problem, don't try to conduct your investigation at arm's length
over the phone. Few vinyl mysteries are unraveled in this manner because critical information is usually communicated badly or not at all. Solving complex problems requires step-
ping on the playing field and conducting a firsthand thorough investigation.

Is it really a problem? Early in the troubleshooting process, assess whether or not the
customer's claim is valid. In spite of the old saying, the customer isn't always right; most
have no idea what is commercially acceptable. Complaints about mottled vinyl, tiny air
bubbles and overlapped seams are really non-issues. Sometimes customers expect vinyl to
do what vinyl wasn't designed to do, such as conform to extreme, compound curves
without seams in the vinyl.

My advice is to listen to the customer, provide an explanation, then move onto the next
topic. Many of these "non-complaints" can be avoided by properly instructing customers

prior to the vinyl application. Mottling on a vinyl surface disappears after a few days in the sun's heat. Tiny air bubbles trapped beneath vinyl eventually breathe out and disappear. In fact, overlapped sections of vinyl graphics are preferred to butt seams.

Keep your cool. Even if some customers react in a loud, insulting and demanding manner, don't fly off the handle. Ask the customer for suggestions on how the complaint might best be handled. In this way, the customer has an opportunity to constructively vent any built-up anger.

Ask better questions. Complaints must be handled in the same systematic manner that a detective uses in solving a crime. This process includes thorough questioning and accurate recording of pertinent information, such as photographs. I've included a list of questions that I hope are helpful when you're looking for trouble.

Asking the right questions is critical. Also, listen carefully and document the answers. (Remember the saying, "God gave you two ears and one mouth. Use them in that proportion.") Improving your listening skills helps your chance of success, and they are essential to good customer relations.

If a customer has a problem with his graphics, he's often defensive and sometimes ready to rumble. Avoid arguing with a customer or implying that he is wrong. In addition, don't dismiss the complaint with a glib explanation. Even if your answer is right, it will seem suspect.

Use a process of elimination. Discovering a problem's root cause usually entails the person investigating the problem by working through his list of potential culprits. Fading colors, for example, can result from various possible causes. Some possibilities include improper material selection, poor processing of inks and clearcoats, repeated chemical spillage and abusive cleaning practices.

In working through your list of suspects, you need to ask detailed questions, carefully record facts, take plenty of photographs and gather the necessary samples and lot numbers of the failed material and the unprocessed raw materials. The lab people at the raw-material manufacturer need samples to conduct thorough testing, which often involves trying to replicate the failure.

Continue your industry education. Education and experience are also indispensable tools in problem solving. The more you know about the business, the easier it will be to determine possible causes of failure. Reading trade publications such as *ST, The Big Picture* and *Screen Printing* should become a daily ritual. More technical information, such as vinyl troubleshooting guides and application guides, is also available from the vinyl manufacturers and industry associations, such as SGIA. Production and technical people can be a source of indispensable knowledge.

Develop a professional network. A technical service manager for one vinyl company recently gave me some sound advice: "No matter how experienced you are, you'll never know it all. The key is to at least know where to get the answers." He meant that you should build a network of industry friends who are experts in the business and can provide necessary information.

The usual suspects

I recently inspected a graphics failure in which a large bubble developed between the plastic sign face and the applied graphics. Because of the sign's complex construction — it consisted of polycarbonate sheet, mounting adhesive, polyester print media, ink, PVC overlaminate and application tape — several possible causes existed. Complicating this mystery was the fact that several manufacturers supplied raw materials.

Another example of a calendered vinyl failure. Is this failure on the part of the material, or the signmaker?

Many of my associates expected the manufacturers to point fingers at each other and then run for cover. While some of this occurred, one manufacturer picked up the tab for all of the materials used in the job. I believe the manufacturer replaced the material to accommodate a good distributor and maintain its good reputation. While this was commendable, it's no substitute for real problem solving.

Vinyl problems result from various causes, which can include the following:

Raw material problems. Every manufacturer occasionally makes something out of spec that slips past quality control. Some problems include adhesive skips, low coating weight, caliper inconsistencies, surface imperfections and compounding problems. These raw material problems can cause adhesive failure, cracking, chalking, poor opacity, excessive shrinkage, fading and color change.

In today's manufacturing environment, these problems are fairly rare. In the last decade, manufacturers have instituted more stringent production standards. As a result, less than 0.5% of a product fails to meet manufacturing specifications.

When raw material causes graphics failure, factors beside manufacturing process could include poor product design, inadequate product testing or grossly exaggerated performance characteristics (sales and marketing departments often do this to the chagrin and consternation of the tech people). If in doubt, call or e-mail the manufacturer's technical department or call a colleague that you trust.

Poor production planning. Many failures occur simply because a signmaker selected the wrong material for the job. In the sign industry, cheaper, calendered vinyl is often substituted for more expensive, translucent films. While calendered films have become better over the years, they still are prone to shrinkage and cracking.

Or, the materials selected for a job could be incompatible. Compatibility of graphic components is critical to the success of any program. The interaction of substrate, adhesive, vinyl, inks, paints, clearcoats and overlaminates involves complex chemistry. Sign professionals are frequently creative in their media combinations. However, combining components without adequate testing can be catastrophic. This brings me to my familiar refrain: Test, don't guess.

In my example of bubbling graphics on a plastic sign face, the print-media manufacturer blamed the mounting adhesive selected from another company. The distributor for the print-media manufacturer maintained the manufacturer was trying to evade the problem.

Still, when it comes to material selection, the ultimate responsibility rests with the sign fabricator. The signshop, screenprinter or digital service provider must determine a raw material's suitability. Many vinyl problems can be traced back to the planning or engineering stage. All too frequently, raw material selection is based on price, not performance.

Fabrication. Just as raw-material manufacturers can encounter process problems, so can vinyl-graphics fabricators. As variables increase in any manufacturing operation, the failure rate grows proportionately.

Look at my example of the plastic sign's failing graphics. Most of the bubbles occurred near the top and bottom edges of the sign face, and very few occurred in the center of the sign. A technical analyst surmised that the laminator's rollers were crowned, creating insufficient pressure towards the edges.

Application problems. Application mistakes account for a high percentage of graphics failures. These mistakes include poor substrate preparation and application techniques. With a plastic sign face, many investigators blamed bubbling on inadequate polycarbonate preparation.

Because polycarbonate sheet absorbs moisture like a sponge, sign fabricators must bake it in an air-circulating oven before decoration. This baking process can last 4-24 hours at 250°F (121°C). It eliminates all of the gasses and moisture, along

Using the right materials, the right tools and the right application technique are essential in fleet graphics application. Failure to use proven materials and reccommended installation tools and procedures resulted in this film failure.

with any residuals (or unreacted polymers still in the sheet). If the polycarbonate isn't baked and prepped properly, the sheet will outgas, causing the applied decorative material to bubble. Also, the residuals that bloom to the surface can contaminate the adhesive, resulting in adhesion failure.

Environmental and customer abuse. Raw-material manufacturers, sign companies and graphics installers aren't always to blame for graphics failures. Some problems result from customer abuse of the graphics.

Years ago, I inspected a failure on 12 painted 4 × 8-ft., plywood construction site signs. All the shriveled vinyl lettering was falling off the surface.

In looking at the peeling letters, we noticed a peculiar clean coating was separating from the vinyl's surface. Further investigation revealed that the builder had his employees varnish all of the signs. The hot solvents in the varnish penetrated the vinyl, attacking the adhesive system.

More common examples of graphics abuse include chemical spillage and improper cleaning. To avoid these problems, the end-user needs to be instructed on the proper care of graphics. Hard-bristle brushes, harsh cleaning chemicals and excessively high-pressure spraying can easily damage vinyl graphics.

Make a list, check it twice
Problem solving doesn't always require a Ph.D. in chemistry, nor expensive, sophisticated laboratory equipment. Finding the solution is often a matter of taking a systematic approach of asking the right questions, careful observation and complete written documentation of the entire procedure, complemented with appropriate photographs and necessary samples.

By using a checklist of questions to guide your investigation, you will present yourself in a professional manner. Here's a sampling of information I would cull from a customer.

- Collect basic information. This includes the customers' contact information, a description of their complaint, and the number of signs, vehicles or other substrates involved.
- Find out who manufactures the substrate or vinyl. Learn the product series, lot number and other pertinent information. If inks, clearcoats or overlaminates have been used, get the manufacturer, series and lot number.
- Descibe the manufacturing method. Were the graphics computer-cut, screenprinted or digitally printed? What finishing operations were involved?
- Obtain particulars about the application itself. Where were the graphics installed, and under what conditions? What was the surface temperature of the substrate at the time of application? How was the substrate cleaned? When was the surface cleaned?
- Know the type of surface. Was the surface smooth, riveted, corrugated or textured? What was the condition of the substrate? Was there any damage to the surface? If it was an old surface, indicate whether the paint was chalked, pitted, peeling, etc.
- Know the substrate's paint history. When was it painted? Who is the paint manufacturer, and what's the product series?
- Find out the application technique. What was the installer's level of experience? What

tools did the installers use? Were the graphics installed wet or dry? If it was a wet application, what type of application fluid was used? Were the graphics edge-sealed? What type of edge sealer was used? Obtain a sample and take plenty of pictures.

- Learn what cleaning and environmental conditions were present. Prior to installation, were the graphics stored, and if so, how long? What were storage conditions? Following the installation, by what method and how frequently were the graphics cleaned? What types of chemicals were used? Were the graphics cleaned using a high-pressure sprayer? Were the graphics subjected to chemical spillage? If they were, find out what kind of chemicals and how often.

Many of the questions that you ask in troubleshooting a complaint are similar to the questions that should be asked during an equipment or site survey conducted prior to starting a graphics project. The right questions often prevent problems.

For those problems that aren't avoided, the best way to evaluate complaints is the "hands-on" approach. Above all, listen to the customer. After examining a complaint and discovering the problem's cause, provide a thorough, effective remedy. Be sure the customer fully understands your explanation by asking questions.

If you don't answer your customer's concerns and complaints fully now, you shouldn't expect repeat business later.

Read the "Vinyl Graphics" column every month in

SIGNS OF THE TIMES

The World Leader in Sign Information Since 1906

Signs of the Times has been the leading magazine of the sign industry since 1906. Thousands of professionals in the signage and graphics industries rely on *Signs of the Times* every month for how-to articles, industry and product comparisons, shop tips, new products and business advice.

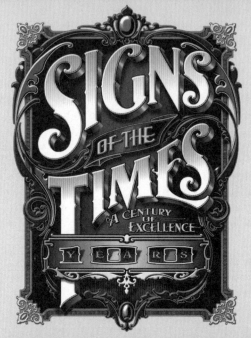

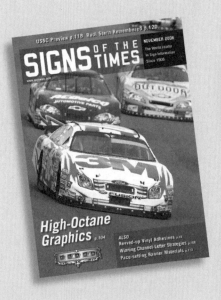